EATCS
Monographs on Theoretical Computer Science
Volume 13

Editors: W. Brauer G. Rozenberg A. Salomaa

Advisory Board: G. Ausiello M. Broy S. Even
J. Hartmanis N. Jones M. Nivat Chr. Papadimitriou
D. Scott

EATCS Monographs on Theoretical Computer Science

Eike Best   César Fernández C.

Nonsequential Processes

A Petri Net View

With 44 Figures

Springer-Verlag
Berlin Heidelberg New York
London Paris Tokyo

Authors

Dr. Eike Best
Prof. César Fernández C.
GMD, Institut für Methodische Grundlagen
Schloß Birlinghoven
D-5205 St. Augustin 1, FRG

Editors

Prof. Dr. Wilfried Brauer
Institut für Informatik, Technische Universität München
Arcisstrasse 21, D-8000 München 2, FRG

Prof. Dr. Grzegorz Rozenberg
Institute of Applied Mathematics and Computer Science
University of Leiden, Niels-Bohr-Weg 1, P.O. Box 9512
NL-2300 RA Leiden, The Netherlands

Prof. Dr. Arto Salomaa
Department of Mathematics, University of Turku
SF-20500 Turku 50, Finland

Library of Congress Cataloging-in-Publication Data.
Best, Eike, 1951- Nonsequential processes : a Petri net view / Eike Best, César Fernández C.
p. cm.–(EATCS monographs on theoretical computer science ; v. 13) Bibliography: p.
Includes index.
ISBN-13: 978-3-642-73485-4   e-ISBN-13: 978-3-642-73483-0
DOI: 10.1007/ 978-3-642-73483-0
1. Petri nets. 2. Machine theory. I. Fernández C., César, 1937-. II. Title. III. Series.
QA267.B48 1988   511.3–dc 19   88-11725 CIP

© Springer-Verlag Berlin Heidelberg 1988
Softcover reprint of the hardcover 1st edition 1988

2145/3140-543210

TO

MONIKA	ORIANA
DAVID	ORIANITA
ROBERT	CÉSAR
SIMON	CARLOS
BENJAMIN	CAROLINA
ANDREAS	BUELI
GUDRUN UND HERBERT	OLGA Y CARLOS

WITH LOVE

Preface

The theory of Petri nets is a part of computer science whose importance is increasingly acknowledged. Many papers and anthologies, whose subject matter is net theory and its applications, have appeared to date. There exist at least seven introductory textbooks on the theory.

The present monograph augments this literature by offering a mathematical treatment of one of the central aspects of net theory: the modelling of concurrency by partially ordered sets. Occurrence nets – which are special nets as well as special partial orders – are proposed by net theory for this purpose. We study both the general properties of occurrence nets and their use in describing the concurrent behaviour of systems.

Occurrence nets may be contrasted with a more language-oriented approach to the modelling of concurrency known as arbitrary interleaving. We will discuss some connections between these two approaches. Other approaches based on partially ordered sets – such as the theory of traces, the theory of event structures and the theory of semiwords – are not considered in this book, in spite of the strong links between them and net theory.

The monograph addresses students in theoretical computer science, mathematicians interested in discrete mathematics, researchers in the area of Petri nets and related topics and professionals in the design of concurrent systems interested in theoretical background. The reader is assumed to have an elementary background in set theory and basic mathematics. Previous knowledge in Petri net theory would be advantageous but is not strictly necessary.

The book is divided into four chapters. Chapter 1 gives the introduction. Chapter 2 deals with general partial order theory and finishes with a specialisation of the theory to posets derived from occurrence nets. Chapter 3 introduces a basic class of Petri nets that may model systems and shows how occurrence nets may be used to represent their nonsequential processes. Chapter 4 shows how properties may be translated from a system to its processes and vice versa.

The logical structure of the book is as follows. Chapter 2 describes a variety of properties of partially ordered sets, ones which may be considered as basic for processes (Sections 2.1 and 2.2) and ones which are interesting for other reasons (Sections 2.3, 2.4 and 2.5). Chapter 3 shows how the behaviour of a system can be defined in terms of occurrence nets. In order to understand this definition, Sections 2.1 and 2.2 are prerequisite reading. In Chapter 4, the properties studied in Sections 2.3, 2.4 and 2.5 are translated into system properties. Hence Sections 2.3, 2.4 and 2.5 (as well as Chapter 3) are required reading in order to understand Chapter 4.

Each chapter is concluded by a number of exercises. Some of them are immediate applications of the concepts defined in the chapter. Some others are extensions of, or recent results in, the theory. The exercises labelled by a star * are harder than the others. An appendix explains the notation used in the book. Another appendix contains a selection of references to the literature and to applications of the theory.

Our appreciation goes to Carl Adam Petri for his willingness to have long discussions about the subject of the book, for his continuous support and, of course, for providing the very subject in the first place. We are indebted to Agathe Merceron and to Raymond Devillers who contributed directly to the material presented in this book. A number of colleagues of our Institute have provided us with comments on all or parts of the manuscript; to all of them we are very grateful. We would also like to thank P.S.Thiagarajan who has given us very useful comments. Our gratitude also goes to two referees who have studied the manuscript carefully and have made valuable suggestions. We are indebted to Elisabeth Münch who has done the drawing of the figures. Last but not least, we thank Grzegorz Rozenberg and Springer-Verlag for making possible the publication of this monograph and for their support.

St.Augustin, February 1988   Eike Best

César Fernández C.

Contents

Chapter 1. Introduction

The concept of *concurrency* plays a fundamental rôle in computer science. A sizeable share of present-day computer systems feature concurrency in an explicit way, being distributed systems or multi-processor systems or both. In the guise of co-existing variables and data structures, concurrency pervades the more conventional sequential systems as well.

A concurrent computer program is normally meant to be executed by more than one processor. It is then possible that two different processors operate independently. Their combined activity is characterised by the property that the subactivities performed by the two processors are linearly ordered in themselves, but not with respect to each other. It may happen that the first processor performs two actions A and C in this order while, independently, the second processor performs some action B, as shown in Figure 1.1. Then B is concurrent to both A and C, but A is not concurrent to C. Consequently, it is reasonable to assume the relation of concurrency to be symmetric, but not necessarily transitive.

Processor 1: ··· A——C ···

Processor 2: ···   B   ···

Fig. 1.1. Concurrent activity

Petri net theory – initiated by Carl Adam Petri in the early Sixties – takes account of this insight from the very start. It constitutes an axiomatic approach to the study of concurrency. One of the assumptions of the theory is that of concurrency being a basic relation which is symmetric but needs not be transitive. In fact, C.A.Petri has suggested to build a theory axiomatically on this assumption only.

A system is modelled in Petri net theory by means of marked nets. The underlying net models the structure of the system, i.e., its local states and their interconnections. Its marking models the initial state of the system. An occurrence rule (also called transition rule or firing rule) determines the behaviour of the system by specifying how the states of the system change under the occurrence of some activity (transition, event) of the system. These ideas may be better understood by considering an example.

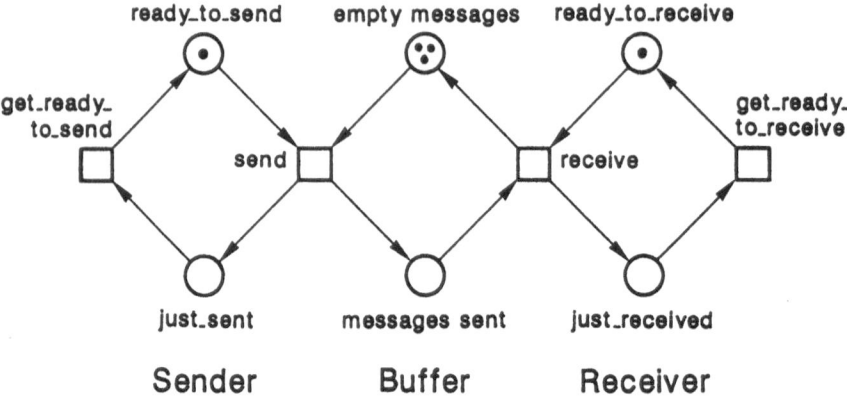

Fig: 1.2. Bounded buffer (initial state)

Let us interpret the Petri net shown in Figure 1.2 as the description of a system, either an existing one or one to be implemented. It contains two sequential activities, a sender and a receiver. The sender consists of two local states (represented by circles), a state *ready_to_send* and a state *just_sent*. Initially, the sender is in the state *ready_to_send*. This is represented pictorially by a token (black dot) on this state. It may perform two activities (represented by squares): *send*, transforming its local state from *ready_to_send* to *just_sent*, and *get_ready_to_send*, transforming its local state from *just_sent* to *ready_to_send*. The receiver is built in an analogous way.

The *send* operation changes the local state of the sender but not that of the receiver. This is expressed by the fact that in Figure 1.2, the *send* operation is not connected to the local states of the receiver. However, the sending also affects the state of the buffer in which messages may be deposited. Our example depicts only the number of messages that have been sent, rather than their contents. Initially there are three empty messages and no produced ones, so that maximally three messages could be produced before any have to be consumed. (The bound of the buffer equals 3.) A *send* operation reduces the number of empty messages by 1 and increases the number of sent messages by 1. A *receive* does the opposite. Figure 1.3 shows the state of the bounded buffer after a single *send* operation.

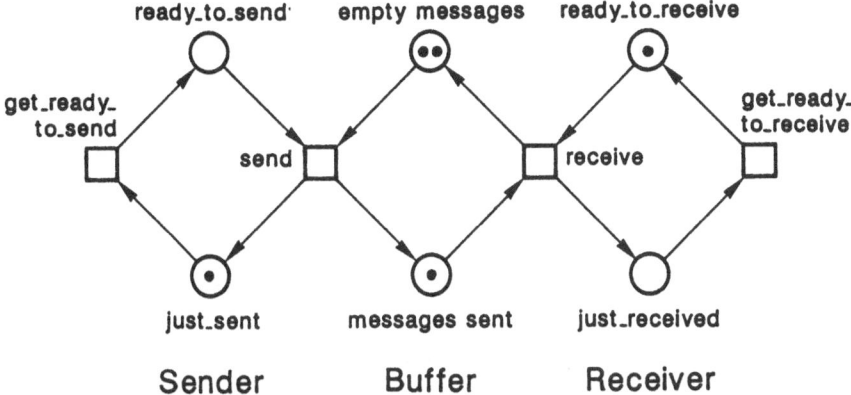

Fig. 1.3. Bounded buffer (after a single *send*)

Let us assume that this system is being executed under the control of two processors, one (the 'S-processor') dedicated to executing the two actions of the sender, the other one (the 'R-processor') dedicated to executing the two actions of the receiver. These two processors might be located at geographically distant sites.

Initially, the R-processor may not execute any action but must wait until at least one message has been produced. Suppose that a message has been produced by the S-processor, i.e., consider the situation after a single *send* shown in Figure 1.3. A *receive* operation could then be executed by the R-processor. Similarly, a *get_ready_to_send* operation could be executed by the S-processor. These two operations are independent of each other, because the latter affects only the local state of the sender while the former affects the receiver's state and the message counting state. Assume now that the following happens: the S-processor executes a *get_ready_to_send* while concurrently, the R-processor executes a *receive* and a *get_ready_to_receive*; the latter two are executed in this order.

The means provided by net theory to describe this complicated-sounding behaviour is a special type of net (called an occurrence net) shown in Figure 1.4.

The activities represented in the occurrence net shown in Figure 1.4 should be thought of as particular occurrences of the actions (and states) shown in the system's description, i.e., in Figure 1.2. For instance, after two actions of the sender, its initial local state *ready_to_send* is again reached; this is shown in the occurrence net by listing the corresponding states twice. The order relations described in the last paragraph are represented formally in this occurrence net. Both the *get_ready_to_send* and the *receive* operations occur after the initial *send*, but they are concurrent (i.e., unordered) with respect to each other. Thus, the concurrent behaviour engendered by the two processors is faithfully modelled by the occurrence net.

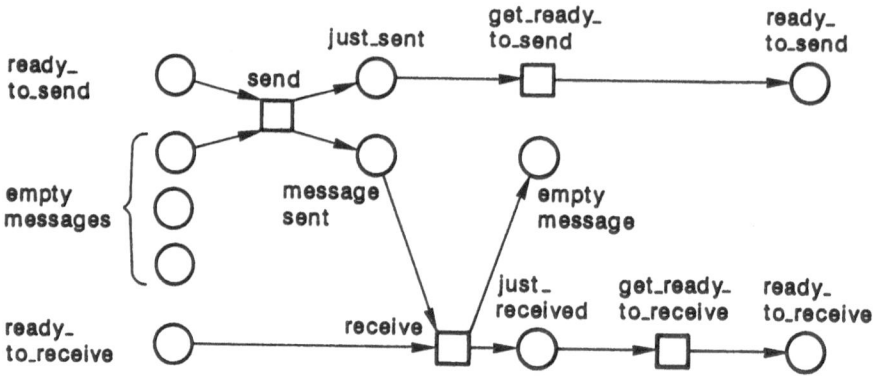

Fig. 1.4. A process (concurrent behaviour) of the bounded buffer

A partially ordered set may be attached in a simple way to an occurrence net. This fact makes the theory of partially ordered sets an adequate mathematical tool to investigate the properties of concurrent processes. This monograph offers a detailed study of the theory of partially ordered sets, focussing on occurrence nets in particular. The bibliographical appendix contains references to several approaches to describe concurrent behaviour by means which differ from occurrence nets. All the same, partially ordered sets play a rôle in almost all of these approaches.

In our presentation, we proceed from the abstract to the concrete. The advantage is that we do not propose an 'ad hoc' theory geared only to specific applications. A risk is that the reader may be overwhelmed by masses of details the motivations of which may not be immediately evident. Nevertheless, we feel that by studying the fine points of partial order theory, the reader will be able to understand better the theory of concurrency.

The detailed structure of the book is as follows:

Chapter 1 – the part you are presently reading – introduces our work.

Chapter 2 is devoted to the investigation of partially ordered sets in general, focussing, towards its end, on occurrence posets – the class of posets associated to occurrence nets.

In Section 2.2 we study a set of discreteness properties the motivation for which is the wish to consider 'discrete models' of behaviour. These properties become relevant in the use of occurrence nets and in the comparison of occurrence nets and arbitrary interleaving.

In Section 2.3 we study two density properties called N-density and K-density, respectively. Their historical motivation is rooted in the fact that a discrete model excludes the ordinary density and hence, one may be interested in investigating any possible alternative density properties which are compatible with discreteness. N-density turns out to be a general requirement for occurrence nets. K-density has a nice 'physical' motivation; as will be shown in Chapter 4, it corresponds to a desirable property of systems.

Section 2.4 is devoted to the study of the concept of D-continuity. Its historical motivation is to transport the properties of the real numbers to posets. Ultimately, the idea is to do calculus on posets. Although we do not achieve this (nor intend to), we still consider the motivation to be strong enough, especially since D-continuity also makes interesting connections as shown in Chapter 4.

In Section 2.5 we study the class of posets which correspond to occurrence nets and which have motivated the whole general study. Occurrence posets are combinatorial and their elements are partitioned into two disjoint sets B and E in such a way that they constitute a record of alternating occurrences of B-elements and E-elements (as illustrated in Figure 1.4 where circles correspond to B-elements and squares correspond to E-elements).

Chapter 3 presents important definitions and results on Petri nets. The central idea is to associate with a Petri net a set of occurrence nets which describes its concurrent behaviour.

Section 3.1 contains basic definitions and results.

Section 3.2 introduces the transition rule on which the definition of the behaviour of a Petri net is based. Occurrence sequences are defined as the arbitrarily interleaved behaviour of marked nets. We also introduce the important notion of safeness.

Section 3.3 contains the definition of the concurrent behaviour of a marked net, i.e., the set of its processes, using occurrence nets. This definition employs some of the concepts of Section 2.2.

Section 3.4 gives an inductive definition of processes. It is shown how a set of processes can be associated with a given occurrence sequence.

A basic class of systems – called systems of finite synchronisation – is defined in Section 3.5; it is shown that for such systems, the relationship between processes and occurrence sequences is weakly invertible.

In Chapter 4 we ask which properties a system must possess so that all of its behaviours satisfy the properties introduced in Chapter 2. We will answer this question for K-density and D-continuity, and it will be demonstrated that both properties correspond to certain desirable (but not automatically always satisfied) properties of systems.

Section 4.2 demonstrates that the property of safeness of a system is equivalent to the K-density of its processes.

In Section 4.3, the notion of a frozen token is introduced. This notion is then related to the D-continuity property defined in Section 2.4, in a similar way as done for safeness and K-density. The absence of frozen tokens is a fairness property regarding the resources of a system; it requires that in an indefinitely long behaviour of the system, none of its resources may be neglected.

Section 4.4, finally, gives a retrospective overview of the book in terms of an elementary class of Petri nets. We specialise the main result of the book to the class of finite 1-safe nets.

The monograph finishes with an appendix containing bibliographical notes and an appendix explaining general notation and terminology. In the former we give a short account of the development of different notions and a list of references about the subject under study, including some references to related work.

Chapter 2. Partially Ordered Sets

2.1 Introduction and Basic Definitions

The present chapter gives some mathematical theory of partially ordered sets. Referring to the appendix on terminology, we recall that a partially ordered set is a pair $(X, \prec)$ where $\prec$ is an irreflexive and transitive relation on X. We shall not immediately give the interpretation of the elements of X. For the purpose of this chapter, it suffices to think of a poset $(X, \prec)$ as describing a history or a process (of a concurrent system) and of an element $x \in X$ as representing a basic occurrence, i.e., an item which has occurred once, and only once, in the history given by $(X, \prec)$. The relation $\prec$ means 'before'; that is, $x \prec y$ means that x has occurred earlier than y in the history given by $(X, \prec)$.

We have been motivated to study partially ordered sets since – as we have seen in the example in the introduction – a behaviour of a system is well represented by a special type of Petri net to whom a poset can be attached. The kind of poset associated to this special type of Petri net is also particular. We think that it is better for the understanding of the subject to study posets in general and then to go to particular cases. However, we shall always keep in mind those particular kinds of posets.

In this section, we will introduce some elementary and important concepts pertaining to partially ordered sets. We shall always assume $X \neq \emptyset$ to exclude the trivial case, and we shall use directed graphs in order to represent posets graphically in the following way: X corresponds to the vertex set of the graph and $\prec$ corresponds to the transitive closure of the relation given by the directed edges (as an example, see Figure 2.1).

We shall use the graphical notation shown in Figure 2.2(i) to indicate that $x \prec y$ and there may be some other elements greater than x and smaller than y. If $x \prec\cdot y$, i.e., there are no elements between x and y, then we always use the graphical notation shown in Figure 2.2(ii). The section on notation and terminology explains further notation and symbols not defined in the main text.

For the rest of this chapter we fix a poset $(X, \prec)$.

The most important relation which can be derived from $(X, \prec)$ is co, denoting unorderedness, definable by x co y iff neither $x \prec y$ nor $y \prec x$; the

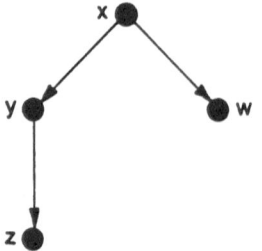

$x \preceq x$, $x \preceq y$, $x \preceq z$, ..., x co x, y co y, y co w, w co z but not y co z, ...

Fig. 2.1. A poset with a non-transitive *co*-relation

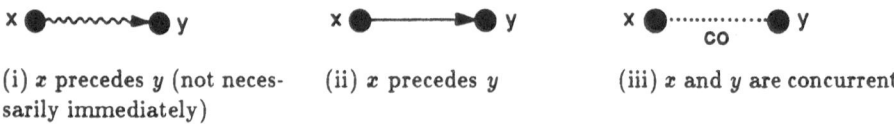

(i) x precedes y (not neces-   (ii) x precedes y (iii) x and y are concurrent
sarily immediately)

Fig. 2.2. Graphical conventions

relation *co* can be interpreted as the concurrency relation (in the history given by $(X, \prec)$) because if neither x occurs earlier than y nor y occurs earlier than x then they occur concurrently. We shall use the graphical notation shown in Figure 2.2(iii) to express x *co* y. A symmetric relation is *li* – denoting orderedness –, defined by x *li* y iff $x \preceq y$ or $y \preceq x$.

We define *li*, *co* and some other related concepts formally as follows:

Definition 2.1.1 [Cuts and Lines]

(i) $li = (\prec \cup \succ \cup id|_X)$, $co = \overline{li} \cup id|_X$.

(ii) $c \subseteq X$ is a co-set (antichain) *iff* $\forall x, y \in c: (x, y) \in co$;
$c \subseteq X$ is a cut *iff* c is a co-set and $\forall z \in X \setminus c \; \exists x \in c: (x, z) \notin co$
(i.e., c is maximal with respect to co);
$C = C(X, \prec)$ is the set of cuts of $(X, \prec)$.

(iii) $l \subseteq X$ is a li-set (chain) *iff* $\forall x, y \in l: (x, y) \in li$;
$l \subseteq X$ is a line *iff* l is a li-set and $\forall z \in X \setminus l \; \exists x \in l: (x, z) \notin li$
(i.e., l is maximal with respect to li);
$L = L(X, \prec)$ is the set of lines of $(X, \prec)$.   □ 2.1.1

As an example, consider the poset shown in Figure 2.3. In this example, we have:

$c_1 = \{z_1, x_3\}$	is a finite cut;
$\{z_1, z_2, z_3\}$	is a co-set which is not a cut;
$\{z_i \mid 1 \le i\}$	is an infinite cut;
$\{x_i \mid 1 \le i\}$	is a li-set which is not a line;
$\{x_i \mid 1 \le i\} \cup \{y_i \mid 1 \le i\}$	is an infinite line;
$\{x_1, x_2, z_1, y_2, y_1\}$	is a finite line.

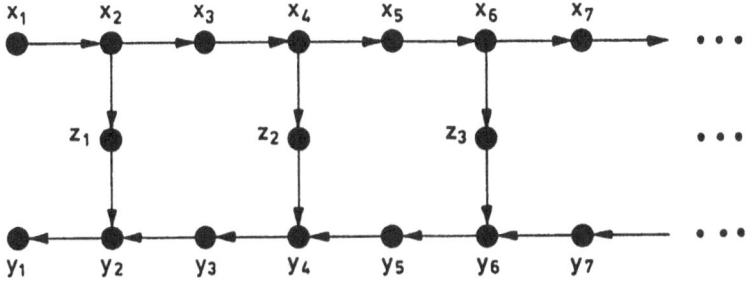

Fig. 2.3. An example illustrating the definition of cuts and lines

Co-sets describe sets of concurrent occurrences. Cuts are maximal co-sets. A cut represents a maximal set of basic occurrences that occur concurrently; in this sense, a cut replaces the notion of a time point. Similarly, li-sets comprise elements of X that occur in sequential order. Lines, being maximal li-sets, may be viewed as the sequential subprocesses of the process described by the poset.

We will assume the axiom of choice to hold, and as a consequence, every co-set c_0 and every li-set l_0 can be extended to a cut $c \supseteq c_0$ and a line $l \supseteq l_0$, respectively.

2.2 Combinatorialness and Discreteness

Partially ordered sets are very general objects. Even linearly ordered sets (such that $co = id$) may be 'discrete' (like the set $\mathbf{Z}$ of integers), dense (like the set $\mathbf{Q}$ of rationals) or uncountable (like the set $\mathbf{R}$ of reals). We aim at a discrete model of processes because information processing systems can generally be regarded as discrete digital systems.

Hence in this section we introduce a few properties which in a certain sense ensure the discreteness of a poset. One of the interesting relations between the basic occurrences of $(X, \prec)$ is that of immediate neighbourship: $x \prec\!\cdot\, y$ if x occurs before y but no other occurrence is between x and y.

Definition 2.2.1 [Immediate Neighbours]

(i)   For $x, y \in X$: $x \prec\!\!\prec y$ *iff* $x \prec y$ and $\neg \exists z \in X : x \prec z \prec y$.

(ii) For $x \in X$: ${}^{\bullet}x = \{y \in X \mid y \prec\!\!\prec x\}$ and $x^{\bullet} = \{y \in X \mid x \prec\!\!\prec y\}$.

(iii) For $Y \subseteq X$, ${}^{\bullet}Y = \bigcup_{y \in Y} {}^{\bullet}y$ and $Y^{\bullet} = \bigcup_{y \in Y} y^{\bullet}$.   □ 2.2.1

Definition 2.2.2 [Density]
$(X, \prec)$ is dense *iff* $\forall x, y \in X : x \prec y \Rightarrow \exists z \in X : x \prec z \prec y$.   □ 2.2.2

To avoid dense (or trivial) posets we should therefore postulate $\prec\!\!\prec \; \neq \emptyset$. Moreover, if $x \prec y$ then, intuitively, y occurs 'some time' after x; it seems reasonable for a discrete model to postulate that between x and y there should be *some* sequence of intermediate immediately neighbouring occurrences. We call this the postulate of combinatorialness.

Definition 2.2.3 [Combinatorialness]
$(X, \prec)$ is combinatorial *iff* $\preceq \; = (\prec\!\!\prec)^{*}$ (or, equivalently, $\prec \; = (\prec\!\!\prec)^{+}$).   □ 2.2.3

Remark 2.2.4
If $(X, \prec)$ is combinatorial and $x \prec y$ then there exists a finite set $\{x_1, x_2, \ldots, x_n\}$ such that $x = x_1 \prec\!\!\prec x_2 \prec\!\!\prec \; \ldots \; \prec\!\!\prec x_n = y$.   □ 2.2.4

Hence, except for the trivial case $li = id|_X$, $(X, \prec)$ cannot be combinatorial and dense at the same time. Thus assuming combinatorialness excludes dense posets such as the rationals or the reals. However, in a certain sense, combinatorialness is still rather weak because if $x \prec y$ then it only postulates the existence of *one* finite $\prec\!\!\prec$-chain between x and y; there may be many other chains between x and y that may be arbitrarily long, even infinitely long. Consider the poset shown in Figure 2.4 which is combinatorial even though the line $\{x_i \mid i \geq 0\} \cup \{y\}$ between x_0 and y is infinite.

This may be an intuitively undesirable situation, since the occurrence of y lies after the infinite chain $x_0 \prec x_1 \prec \ldots$ We exclude it by means of the following definition.

Definition 2.2.5 [Weak discreteness]
$(X, \prec)$ is weakly discrete *iff* $\forall x, y \in X \; \forall l \in L : |[\,x, y\,] \cap l| \in \mathbf{N}$.   □ 2.2.5

Weak discreteness disallows infinite chains between any two elements but does not disallow infinite chains altogether. For example, $(\mathbf{Z}, <)$ is weakly discrete with an infinite li-set $\mathbf{Z}$, where $<$ is the usual ordering on the integers. Our first theorem states that weak discreteness is a stronger property than combinatorialness.

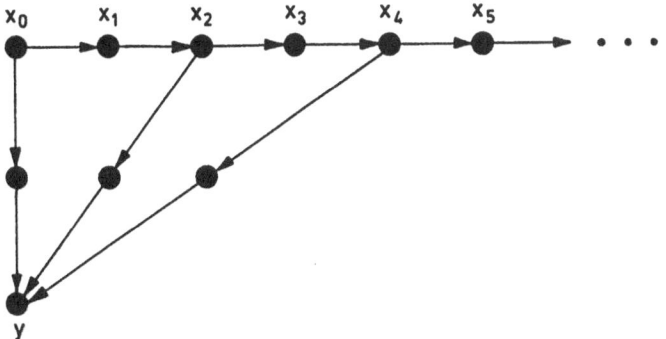

Fig. 2.4. A poset which is combinatorial but not weakly discrete

Theorem 2.2.6 [Weak discreteness implies combinatorialness]
If $(X, \prec)$ is weakly discrete then it is combinatorial.

Proof: The proof of $(\prec)^* \subseteq \preceq$ is immediate; hence we only have to prove that $\preceq \subseteq (\prec)^*$.

Let $(x, y) \in \preceq$, i.e., $x \preceq y$. Let $l \in L$ such that $x, y \in l$. Since $(X, \prec)$ is weakly discrete, there are $x_1, x_2, \ldots, x_n \in X$ such that $x = x_1$, $x_n = y$ and

$$[x, y] \cap l = \{x_1, x_2, \ldots, x_n\}.$$

Since $x_1, x_2, \ldots, x_n \in l$, we can assume, without loss of generality, that

$$x_1 \prec x_2 \prec x_3 \prec \ldots x_i \prec x_{i+1} \prec \ldots \prec x_n.$$

Assume $\exists z \in X$ such that $x_i \preceq z \preceq x_{i+1}$. In this case, it is easy to see that $z \in l$ and $z \in [x, y]$, i.e., $z \in [x, y] \cap l$. This means that $z = x_i$ or $z = x_{i+1}$, i.e.,

$$x_i \prec\!\!\prec x_{i+1} \text{ for all } i = 1, 2, 3, \ldots, n - 1.$$

It follows, then, that $(x, y) \in (\prec\!\!\prec)^{n-1}$, i.e., $(x, y) \in (\prec\!\!\prec)^*$.

In all: $\preceq \subseteq (\prec\!\!\prec)^*$, which accomplishes the proof.   □ 2.2.6

As far as 'infinite pasts' and 'infinite futures' are concerned, weak discreteness makes a poset look a bit like $(\mathbf{Z}, <)$, that is: no single occurrence may lie 'after' an infinite future, and likewise, no single occurrence may lie 'before' an infinite past. However, there may still be posets which are infinite with respect to their 'breadth' which pose intuitive problems, such as the one shown in Figure 2.5.

In Figure 2.5, all chains between x and y are finite, but there is no bound on their length that depends only on x and y. The next definition introduces a discreteness condition designed to exclude this possibility.

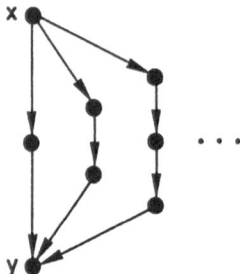

Fig. 2.5. A poset which is weakly discrete but not boundedly discrete

Definition 2.2.7 [Bounded discreteness]
$(X, \prec)$ is boundedly discrete *iff* $\forall x, y \in X \ \exists n \in \mathbf{N}^+ \ \forall l \in L: |[x, y] \cap l| < n.$
$\square$ 2.2.7

Clearly, bounded discreteness is stronger than weak discreteness. The next theorem states that unless some occurrence has infinitely many immediate predecessors or successors (a case which we will later have reason to exclude), weak discreteness and bounded discreteness are the same. In fact, both are equivalent to a property called interval-finiteness. We will define and state these connections next.

Definition 2.2.8 [Degree-finiteness]
$(X, \prec)$ is of finite degree *iff* $\forall x \in X: |{}^\bullet x| \in \mathbf{N} \wedge |x^\bullet| \in \mathbf{N}.$
$\square$ 2.2.8

Definition 2.2.9 [Interval-finiteness]
$(X, \prec)$ is interval-finite *iff* $\forall x, y \in X: |[x, y]| \in \mathbf{N}.$
$\square$ 2.2.9

As an example, reconsider the poset shown in Figure 2.3; it is not boundedly discrete, not weakly discrete and not interval-finite, but it is of finite degree.

Theorem 2.2.10 [Discreteness and interval-finiteness]

(i)   *If $(X, \prec)$ is interval-finite then it is boundedly discrete.*

(ii)   *If $(X, \prec)$ is boundedly discrete then it is weakly discrete.*

(iii) *If $(X, \prec)$ is weakly discrete and of finite degree then it is interval-finite.*

Proof:

(i)   Obvious from Definitions 2.2.9 and 2.2.7.

(ii)   Obvious from Definitions 2.2.7 and 2.2.5.

(iii)   We are going to show that $|[x,y]| \in \mathbf{N}$ for all $x, y \in X$.

If $x \text{ co } y$ or $x \succ y$ then this is trivially true. We consider the case $x \prec y$. Assume that $|[x,y]| \notin \mathbf{N}$. We know that $(X, \prec)$ is weakly discrete and hence combinatorial by Theorem 2.2.6, and then if $x \prec z$ we have $\exists n \in \mathbf{N}^+$ such that $x(\prec)^n z$, that is, $\exists z_1, z_2, \ldots, z_{n-1}$ such that

$$x \prec z_1 \prec z_2 \prec \ldots \prec z_{n-1} \prec z;$$

in particular, $z_1 \in x^\bullet$. Because of this, by degree-finiteness and because $x \prec y$, $|x^\bullet| \in \mathbf{N}^+$; then

$$x^\bullet = \{y_1^1, y_1^2, \ldots, y_1^{k_1}\}$$

with $k_1 \geq 1$. Using this fact it follows immediately that

$$[x,y] = \{x\} \cup \bigcup_{j=1}^{k_1} [y_1^j, y].$$

Now, $|[x,y]| \notin \mathbf{N}$ implies that $|\bigcup_{j=1}^{k_1} [y_1^j, y]| \notin \mathbf{N}$, and the latter implies that $\exists y_1^i, 1 \leq i \leq k_1$ such that $|[y_1^i, y]| \notin \mathbf{N}$.

Put $y_1^i = x_1$. Then we know: $x_1 \in x^\bullet$, $x_1 \prec y$ and $|[x_1, y]| \notin \mathbf{N}$. We can repeat the process to find an x_2 such that $x_2 \in x_1^\bullet$, $x_2 \prec y$ and $|[x_2, y]| \notin \mathbf{N}$. In fact, we can construct a set $U = \{x_1, x_2, \ldots\}$ such that: $x_k \in x_{k-1}^\bullet$, $x_k \prec y$ and $|[x_k, y]| \notin \mathbf{N}$ for $1 \leq k$.

Clearly $|U| \notin \mathbf{N}$, $U \subseteq [x,y]$ and $\{x,y\} \cup U$ is a li-set.

Let $l \in L$ be such that $l \supseteq \{x,y\} \cup U$.

Now $l \cap [x,y] \supseteq U$, hence $|l \cap [x,y]| \geq |U| \notin \mathbf{N}$, hence $|l \cap [x,y]| \notin \mathbf{N}$. Hence $(X, \prec)$ is not weakly discrete, a contradiction.

Hence $|[x,y]| \in \mathbf{N}$, and the proof is done.   □ 2.2.10

It is usually the case that a discrete system starts its operations in a certain 'state' that may correspond to a certain cut in the process associated to these operations. It is then appropriate to require that none of the subsequent occurrences in the poset has an infinite distance to that cut. This idea is captured by the next definition.

Definition 2.2.11 [Discreteness with respect to a cut]
Let $c \in C(X, \prec)$. $(X, \prec)$ is discrete with respect to c *iff*
$$\forall x \in X \ \exists n \in \mathbf{N}^+ \ \forall l \in L(X, \prec): |[c,x] \cap l| < n \wedge |[x,c] \cap l| < n. \quad □ \ 2.2.11$$

There is an analogue of Theorem 2.2.10(iii), putting discreteness with respect to c in place of weak discreteness. We state this result next.

Theorem 2.2.12 [Discreteness w.r.t. a cut and the finiteness of cut-intervals]
Suppose that the poset $(X, \prec)$ is discrete with respect to a cut c and of finite degree. Then for all $x \in X$: $[x, c] \cup [c, x]$ is a finite set.

Proof: If $x \in c$ then $[x, c] \cup [c, x] = \{x\}$ and the result is true. If $x \notin c$ then either $x \prec y$ for some $y \in c$ or $y \prec x$ for some $y \in c$ because c is a cut.

Assume $x \prec y$ for some $y \in c$; the other case can be proved analogously. Then $[c, x] = \emptyset$ because c is a co-set, and we have to prove that $[x, c]$ is a finite set. Assume the contrary; we will derive a contradiction. As in the proof of Theorem 2.2.10(iii), using degree-finiteness, we may construct an infinite li-set

$$x_1 \prec x_2 \prec x_3 \prec \ \ldots$$

such that $x = x_1$ and $\forall i \geq 1 \ \exists y_i \in c : x_i \prec y_i$. This contradicts discreteness with respect to c because for arbitrary $n \in \mathbf{N}^+$, any line $l \supseteq \{x_1, \ldots, x_n, y_n\}$ satisfies $|[x, c] \cap l| \geq n$. $\qquad \square$ 2.2.12

Remark 2.2.13 [Discreteness w.r.t. a cut implies bounded discreteness]
If there exists a cut c such that $(X, \prec)$ is discrete with respect to c then $(X, \prec)$ is boundedly discrete. We mention this fact without proving it, since its proof will follow indirectly using some of the theorems of the remainder of this section (see Figure 2.7). $\qquad \square$ 2.2.13

Bounded discreteness, interval-finiteness and discreteness with respect to a cut c all make a connection to a certain notion of observability. The idea is to postulate that an observer exists which 'has a linear discrete time scale'. A linear discrete time scale is, by definition, simply the ordered set $(\mathbf{Z}, <)$ of integers. The requirement that such an observer exists for $(X, \prec)$ can be taken to mean that $(X, \prec)$ is embeddable into $(\mathbf{Z}, <)$ in one of the following senses:

Definition 2.2.14 [Observability]

(i) $(X, \prec)$ is observable *iff* $\exists f : X \to \mathbf{Z} : \forall x, y \in X : x \prec y \Rightarrow f(x) < f(y)$; f is called an observer of $(X, \prec)$.

(ii) $(X, \prec)$ is injectively observable *iff* $(X, \prec)$ is observable with an injective observer f.

(iii) $(X, \prec)$ is observable with respect to a cut c of $(X, \prec)$ *iff* $(X, \prec)$ is observable with an observer $f : X \to \mathbf{Z}$ such that $\forall x \in c : f(x) = 0$. $\qquad \square$ 2.2.14

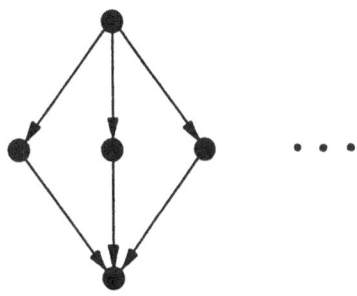 ...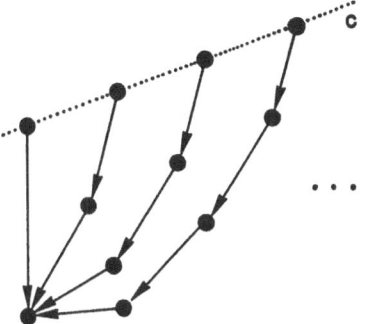

(i) Discrete and observable w.r.t. c for all c, but neither interval-finite nor injectively observable

(ii) Boundedly discrete, interval-finite and injectively observable, but neither discrete w.r.t. c nor observable w.r.t. c

Fig. 2.6. Observability and discreteness

For instance, $(\mathbf{Z}, <)$ is trivially observable while the poset shown in Figure 2.5 is not. Clearly, observability is strictly weaker than both injective observability and observability with respect to a cut. Figure 2.6 shows that the latter are in general incomparable.

The next theorems give the connection between the observability notions defined in Definition 2.2.14 and the discreteness properties introduced prior to that.

Theorem 2.2.15 [Observability and bounded discreteness]

(i) *If a poset $(X, \prec)$ is observable then it is boundedly discrete.*

(ii) *If X is countable and $(X, \prec)$ is boundedly discrete then $(X, \prec)$ is observable.*

Proof:

(i)   Let $x, y \in X$. We have to show that there exists a number $n \in \mathbf{N}^+$ such that $\forall l \in L : |[x, y] \cap l| < n$.

If $x \ co \ y$ or $x \succ y$ then the result is trivially true. We can assume that $x \prec y$. Since $(X, \prec)$ is observable we get $f(x) < f(y)$ where $f : X \to \mathbf{Z}$ is an observer of $(X, \prec)$. Let $n = f(y) - f(x) + 2 \in \mathbf{N}^+$.

Claim: $\forall l \in L : |[x, y] \cap l| < n$.

We conduct an indirect proof of this claim. Assume $\exists l' \in L$ such that $|[x, y] \cap l'| \geq n$. Then there exist at least n different elements $x_1, x_2, \ldots, x_n$ in $[x, y] \cap l'$. Since $\{x_1, x_2, \ldots, x_n\} \subseteq [x, y] \cap l'$ is a finite li-set, we can order it (re-enumerating if necessary) in such a way that

$$x \preceq x_1 \prec x_2 \prec x_3 \prec \ldots \prec x_n \preceq y.$$

Since $(X, \prec)$ is observable we get:

$$f(x) \leq f(x_1) < f(x_2) < f(x_3) < \ldots < f(x_n) \leq f(y),$$

which means that we have at least $n - 2 = f(y) - f(x)$ different increasing integers strictly between the integers $f(x)$ and $f(y)$, which is clearly impossible. So, assuming $\exists l' \in L : |[x, y] \cap l'| \geq n$ we get a contradiction, and this achieves the proof.

(ii)   Because the poset is countable, there exists an enumeration

$$X = \{x_0, x_1, x_2, \ldots\}$$

of the set X. For $x, y \in X$ we define the number $\nu(x, y)$ by

$$\nu(x, y) = \max_{l \in L} \{|[x, y] \cap l|\}.$$

By bounded discreteness, we know that the maximum exists, i.e., $\nu(x, y) \in \mathbf{N}$. We need two auxiliary sets, defined for all integers $i \geq 0$:

$$\alpha_i = \{k \mid 0 \leq k < i \wedge x_k \prec x_i\}$$
$$\beta_i = \{j \mid 0 \leq j < i \wedge x_i \prec x_j\}.$$

α_i (β_i) is the set of indices before i of such elements which are before x_i (after x_i, respectively); clearly, $\alpha_0 = \beta_0 = \emptyset$. Using these definitions we may now define a suitable observer $f : X \to \mathbf{Z}$ inductively (on $i \geq 0$) as follows:

$$f(x_i) = \begin{cases} \max_{k \in \alpha_i} \{f(x_k) + \nu(x_k, x_i) - 1\} & \text{if } \alpha_i \neq \emptyset \\ \min_{j \in \beta_i} \{f(x_j) - \nu(x_i, x_j) + 1\} & \text{if } \alpha_i = \emptyset \wedge \beta_i \neq \emptyset \\ 0 & \text{if } \alpha_i = \beta_i = \emptyset \end{cases}$$

We claim that f, so defined, is an observer. To prove this claim, we shall prove, by induction on $i \geq 0$, the stronger claim that:

$$\forall i \geq 0 \; \forall k, j : 0 \leq k, j < i \wedge x_k \prec x_j \Rightarrow f(x_j) - f(x_k) \geq \nu(x_k, x_j) - 1.$$

This formula means that the difference in terms of f-values between x_k and x_j always equals or exceeds the maximal distance between x_k and x_j. The fact that f is an observer follows immediately from this formula. To prove the formula, we proceed by induction on i.

$\underline{i = 0}$:   The formula is trivially true.

$\underline{\{0, \ldots, i\} \to i + 1}$:   We have to prove that

$$f(x_j) - f(x_k) \geq \nu(x_k, x_j) - 1$$

whenever $0 \leq k < i + 1$, $0 \leq j < i + 1$ and $x_k \prec x_j$. If both $j < i$ and $k < i$ then the inequality follows immediately from the induction hypothesis. By $x_k \prec x_j$, we cannot have both $k = i$ and $j = i$. Hence two cases remain to be considered: $0 \leq k < j \wedge j = i$ (Case 1) and $0 \leq j < k \wedge k = i$ (Case 2).

Case 1:   Suppose $x_k \prec x_i$ with $0 \leq k < i$; we have to prove that

$$f(x_i) - f(x_k) \geq \nu(x_k, x_i) - 1.$$

But $\alpha_i \neq \emptyset$ since $k \in \alpha_i$; hence the first clause of the definition of f applies, and hence we have:

$$f(x_i) \geq f(x_k) + \nu(x_k, x_i) - 1,$$

from which the desired relation follows.

Case 2:   Suppose $x_i \prec x_j$ with $0 \leq j < i$; we have to prove that

$$f(x_j) - f(x_i) \geq \nu(x_i, x_j) - 1.$$

But $\beta_i \neq \emptyset$ because $j \in \beta_i$.

Case 2a: $\alpha_i = \emptyset$.
Then the second clause of the definition of f applies. Hence we have:

$$f(x_i) \leq f(x_j) - \nu(x_i, x_j) + 1,$$

from which the desired relation follows.

Case 2b: $\alpha_i \neq \emptyset$.
Then the first clause of the definition of f applies. Let $k' \in \alpha_i$ be such that

$$f(x_i) = f(x_{k'}) + \nu(x_{k'}, x_i) - 1$$

(i.e., such that the maximum given in the first clause is assumed; such an index k' exists because α_i is a finite set). Then we have $x_{k'} \prec x_i \prec x_j$ and, clearly, also

$$\nu(x_{k'}, x_j) \geq \nu(x_{k'}, x_i) + \nu(x_i, x_j) - 1.$$

Then:

$$
\begin{aligned}
f(x_i) &= f(x_{k'}) + \nu(x_{k'}, x_i) - 1 \\
&\leq f(x_{k'}) + \nu(x_{k'}, x_j) - \nu(x_i, x_j) + 1 - 1 \\
&\quad \text{(because of the inequality just mentioned)} \\
&\leq f(x_j) - \nu(x_{k'}, x_j) + 1 + \nu(x_{k'}, x_j) - \nu(x_i, x_j) \\
&\quad \text{(by the induction hypothesis for } k', j) \\
&= f(x_j) - \nu(x_i, x_j) + 1,
\end{aligned}
$$

from which the desired relation follows.   □ 2.2.15

There exist examples which show that for uncountable X, bounded discreteness does not imply observability (see Exercise 11).

Theorem 2.2.16 [Injective observability and interval-finiteness]
A poset $(X, \prec)$ is injectively observable iff it is interval-finite and countable.

Proof:

($\Rightarrow$)   The fact that X is countable follows trivially since $f: X \rightarrow \mathbf{Z}$ is an injection from X into $\mathbf{Z}$. Let us prove, now, that X is interval-finite.

Let $x, y \in X$. Then $|[x, y]| = |f([x, y])|$ since f is injective. Clearly, also,

$$f([x, y]) \subseteq [f(x), f(y)] \subseteq \mathbf{Z}$$

since f is an observer. Then

$$|[x, y]| = |f([x, y])| \leq |[f(x), f(y)]| \in \mathbf{N},$$

i.e., $|[x, y]| \in \mathbf{N}$, and the proof is done.

($\Leftarrow$)   Let $(X, \prec)$ be interval-finite and countable. Then there exists an enumeration $X = \{x_0, x_1, x_2, x_3, \ldots\}$. Define

$$A_i = \uparrow\{x_0, \ldots, x_i\} \cap \downarrow\{x_0, \ldots, x_i\}.$$

We have:

$$X = \bigcup_{i=0}^{\infty} A_i.$$

By interval-finiteness, the sets A_i are finite. Furthermore, $A_0 = \{x_0\}$ and $A_i \subseteq A_{i+1}$ for all $i \geq 0$. Now we define inductively a sequence $f_0, f_1, f_2, \ldots$ of injective observers of $A_0, A_1, A_2, \ldots$ Let $f_0: A_0 \rightarrow \mathbf{Z}$ be any function; f_0 is clearly an injective observer of A_0.

Claim: Let $f_i: A_i \rightarrow \mathbf{Z}$ be an injective observer of A_i. Then f_i can be extended to an injective observer f_{i+1} on A_{i+1}.

Proof: Let A be any finite set such that $A_i \subseteq A \subseteq X$ and $A = \uparrow A \cap \downarrow A$. We prove, by induction on $|A|$, that f_i can be extended to an injective observer f of A.

If $|A| = |A_i|$ then $A = A_i$, and $f = f_i$ is already an injective observer of A.

Suppose that $|A| > |A_i|$. If $A_i \supseteq Min(A) \cup Max(A)$ then $A_i = A$ (since $A_i \subseteq A$ by supposition and $A \subseteq A_i$ because A is finite, $A = \uparrow A \cap \downarrow A$ and $A_i = \uparrow A_i \cap \downarrow A_i$), contradicting $|A| > |A_i|$. Hence $\exists y \in Min(A) \cup Max(A)$ such that $y \notin A_i$. Assume that $y \in Min(A)$ and consider $A' = A \setminus \{y\}$. Since $y \notin A_i$, we have $A_i \subseteq A'$. Furthermore, clearly, $A' = \uparrow A' \cap \downarrow A'$. By the induction hypothesis, f_i can be extended to an injective observer f' of A'. Now define $f: A \rightarrow \mathbf{Z}$ by:

$$f(x) = f'(x) \text{ for } x \in A' \text{ and}$$
$$f(y) = \min_{x \in A'}\{f'(x)\} - 1.$$

Clearly, f is an injective observer on A. If $y \in Max(A)$ then the proof is similar.

Put $A = A_{i+1}$ to obtain the desired observer f_{i+1} of A_{i+1}.   □ *Claim*

Finally, define $f: X = \bigcup_{i=0}^{\infty} A_i \to \mathbf{Z}$ by

$$f(x) = f_i(x) \quad \text{if} \quad x = x_i$$

(then $f(x) = f_j(x)$ for all j such that $x \in dom(f_j)$). It is easy to check that f is an injective observer of $(X, \prec)$.   □ 2.2.16

Theorem 2.2.17 [Observability and discreteness w.r.t. a cut c]
Let $(X, \prec)$ be a poset and $c \in C(X, \prec)$.
$(X, \prec)$ is observable with respect to c iff $(X, \prec)$ is discrete with respect to c.

Proof: Let $f: X \to \mathbf{Z}$ be an observer such that $f(x) = 0$ for each $x \in c$. Let $n(x) = |f(x)| + 1$. Then it is not difficult to see that the condition of discreteness with respect to c is satisfied.

Conversely, if $(X, \prec)$ is discrete with respect to c then we may define an observer $f: X \to \mathbf{Z}$ by putting, for each $x \in X$:

$$f(x) = \begin{cases} -\max_{l \in L}\{|[x,c] \cap l|\} & \text{if } \exists z \in c : x \prec z \\ 0 & \text{if } x \in c \\ \max_{l \in L}\{|[c,x] \cap l|\} & \text{if } \exists z \in c : z \prec x \end{cases}$$

f is well-defined because of the discreteness with respect to c, and it is routine to check that f is an observer.   □ 2.2.17

This concludes the definition of a hierarchy of discreteness properties which are motivated by the wish to exclude, with a view to a discrete process model, unreasonably 'dense' posets. The hierarchy and the connections are summarised in Figure 2.7.

2.3 N-density and K-density

In this section we study two density properties called N-density and K-density, respectively. To motivate N-density, we use an example from computer science. Imagine a sequential process that sends a message to its right neighbour and then receives a message from its left neighbour. It may happen that the state in which the left neighbour has not yet sent its message is concurrent with the state in which the right neighbour has already received its message. In this case, the middle process must be in a state that is *after* the sending of the right hand message but *before* the receiving of the left hand message. The existence of such a state is postulated by N-density.

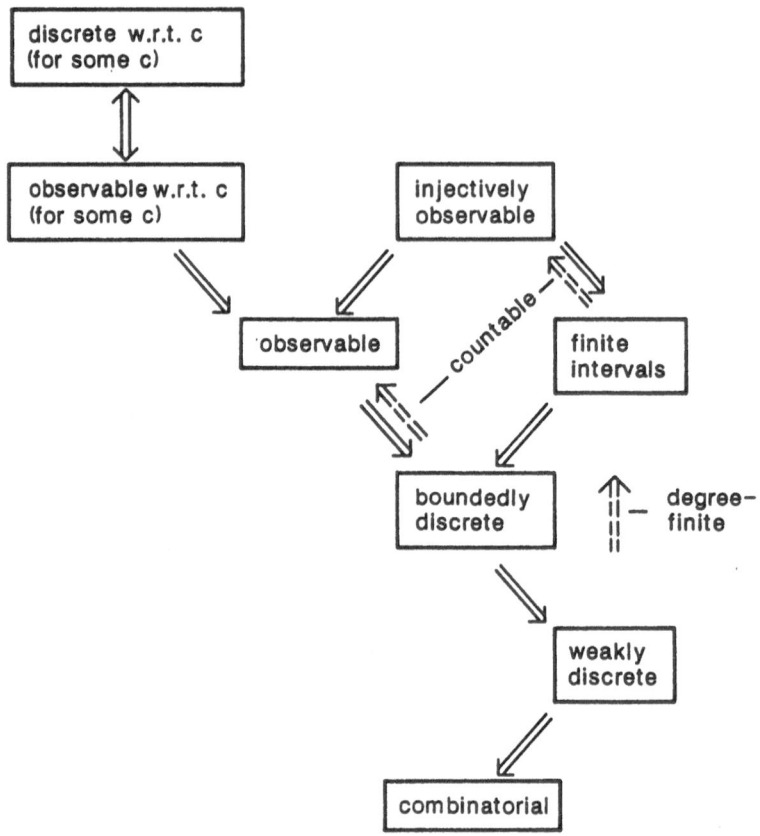

Fig. 2.7. Summary of Section 2.2

In other words, N-density postulates an occurrence between two basic occurrences x and y with $x \prec y$, provided that there could be a 'time point' between x and y. Since the notion of a time point is replaced by that of a cut, we consider the case that there exist occurrences x' and y' such that $x \prec x'$, $y' \prec y$ and x co y' co x' co y. In that case, any cut through x' and y' lies between x and y, and an occurrence z between x and y is postulated by N-density.

Definition 2.3.1 [N-density]
$(X, \prec)$ is N-dense *iff* for all $x, y, x', y' \in X$:
$x \prec y \ \wedge \ x \prec x' \ \wedge \ y' \prec y \ \wedge \ (x \text{ co } y' \text{ co } x' \text{ co } y) \Rightarrow$
$\quad \exists z \in X \colon x \prec z \prec y \ \wedge \ (x' \text{ co } z \text{ co } y').$ □ 2.3.1

The 'N' stems from the shape of the diagram in Figure 2.8 that may be used to explain the property. Figure 2.9 shows a poset which is not N-dense.

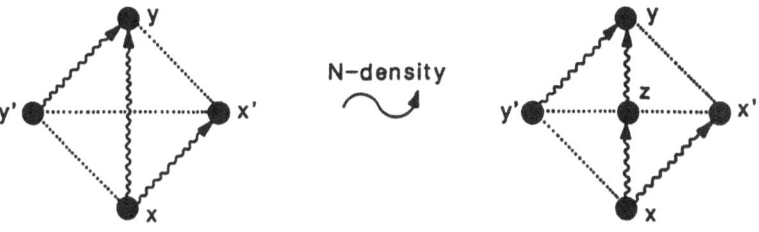

Fig. 2.8. Illustration of N-density

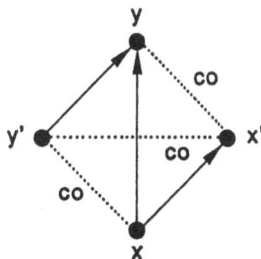

Fig. 2.9. A poset which is neither N-dense nor K-dense

As opposed to density, N-density is compatible with all of the discreteness properties introduced in Section 2.2. Indeed, a poset can easily, say, be interval-finite and N-dense at the same time. N-density can be strengthened without losing that property. The K-density property (where the 'K' stands for 'Kombinatorisch' in German) introduced next provides such a strengthening. It is based on the idea of interpreting cuts as the states and lines as the sequential subprocesses of the process described by $(X, \prec)$. It is always true that $|c \cap l| \leq 1$ for a cut c and a line l. K-density postulates $c \cap l \neq \emptyset$ (i.e., $|c \cap l| = 1$), meaning that every basic occurrence of a subprocess l must be in a definitive state.

Definition 2.3.2 [K-density]
$(X, \prec)$ is K-dense *iff* $\forall c \in C \; \forall l \in L: c \cap l \neq \emptyset$.   □ 2.3.2

Figure 2.9 shows a poset which is neither K-dense nor N-dense. A simple relation between K-density and N-density is given by the next proposition.

Proposition 2.3.3 [K-density implies N-density]
If $(X, \prec)$ is K-dense then it is N-dense.

Proof: If $x, y, x'y'$ satisfy the conditions of Definition 2.3.1 then any cut through $\{x', y'\}$ meets any line through $\{x, y\}$ at an element z with the required properties.   □ 2.3.3

The converse is not true, as is proved by the three posets of Figure 2.10 which are N-dense but not K-dense.

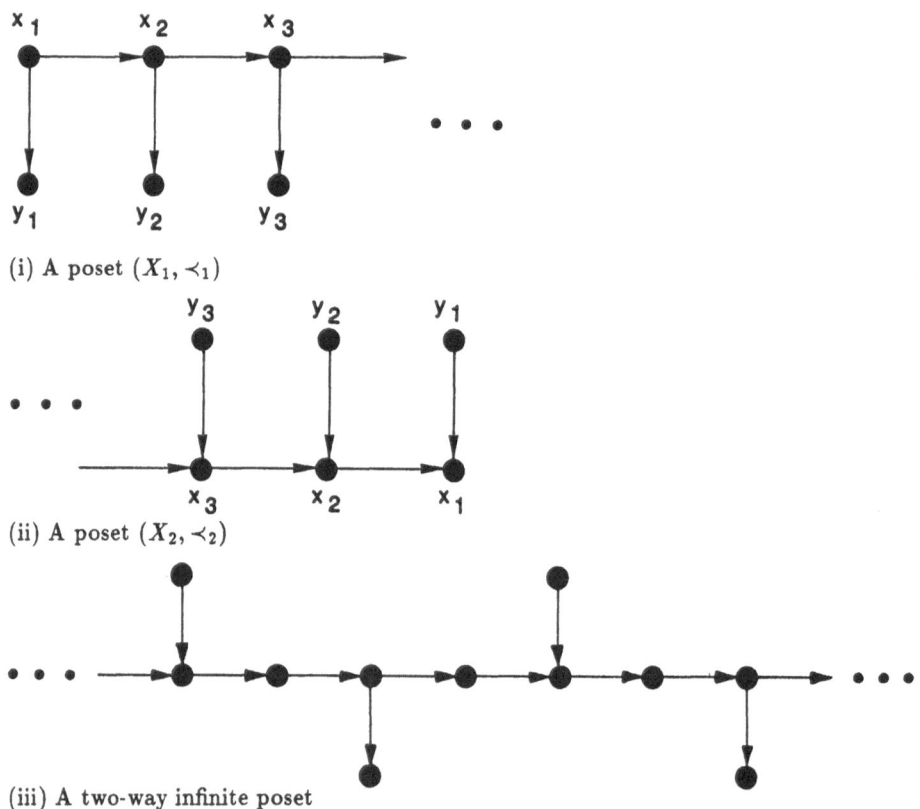

(i) A poset $(X_1, \prec_1)$

(ii) A poset $(X_2, \prec_2)$

(iii) A two-way infinite poset

Fig. 2.10. Posets which are N-dense but not K-dense

In the following we describe a characterisation of K-density. This result shows that under certain assumptions, K-density can be related to the property of weak discreteness discussed in the previous section and, furthermore, can be characterised by the absence of the substructures shown in Figure 2.10(i,ii). To state the result, we need two auxiliary definitions.

Definition 2.3.4 [Line crossing property]
A poset $(X, \prec)$ has the line crossing property *iff*
$\forall c \in C \; \forall l \in L: \downarrow c \cap l \neq \emptyset$ and $\uparrow c \cap l \neq \emptyset$.   □ 2.3.4

This means that a line may not stay properly below or above a cut. Notice that the line crossing property is implied by K-density and fails to hold in the examples of Figure 2.10.

Definition 2.3.5 [Embeddability]
A poset $(X', \prec')$ is embeddable into $(X, \prec)$ *iff* there exists an injection $\gamma: X' \to X$ such that $\forall x, y \in X': x \prec' y \iff \gamma(x) \prec \gamma(y)$.   □ 2.3.5

Theorem 2.3.6 [Characterisation of K-density]
Let $(X, \prec)$ be N-dense and combinatorial.
Then the following three properties are equivalent:

(i)   *$(X, \prec)$ is K-dense.*

(ii) *$(X, \prec)$ is weakly discrete and has the line crossing property.*

(iii) *None of the two posets shown in Figure 2.10(i,ii) is embeddable into $(X, \prec)$.*

In order to make the proof of the theorem more easily accessible to the reader, we are going to split it into three propositions following the sequence

(i) $\Rightarrow$ (ii) $\Rightarrow$ (iii) $\Rightarrow$ (i).

We assume, in the following three propositions, that $(X, \prec)$ is N-dense and combinatorial.

Proposition 2.3.7
If $(X, \prec)$ is K-dense then it is weakly discrete and has the line crossing property.

Proof: K-density implies trivially the line crossing property.
We prove now that: $(X, \prec)$ is K-dense $\Rightarrow$ $(X, \prec)$ is weakly discrete. The proof is an indirect one. Assume that $(X, \prec)$ is not weakly discrete. Then it is easy to see that $\exists x_0, y_0 \in X \; \exists l_0 \in L$ such that $x_0, y_0 \in l$ and $|l_0 \cap [x_0, y_0]| \notin \mathbf{N}$. Define:

$$l_1 = l_0 \cap [x_0, y_0];$$
$$X_1 = \{x \in l_1 \mid |l_1 \cap \downarrow x| \in \mathbf{N}^+\};$$
$$Y_1 = l_1 \setminus X_1.$$

Clearly, $x_0 \in X_1$ and $y_0 \in Y_1$, i.e., $X_1 \neq \emptyset$ and $Y_1 \neq \emptyset$. Furthermore, if $Max(X_1) \neq \emptyset$ and $Min(Y_1) \neq \emptyset$ then for $y \in Min(Y_1)$ we have $|l_1 \cap \downarrow y| \in \mathbf{N}^+$ which is a contradiction with the definition of Y_1. This means that either $Max(X_1) = \emptyset$ or $Min(Y_1) = \emptyset$ or both.

Case 1: $Max(X_1) = \emptyset$.
Define:

$$c_1 = \{x \in X \mid \exists y \in \downarrow X_1 : y \prec x\} \setminus (\downarrow X_1).$$

The idea of the proof is:

(1)   Prove that c_1 is a co-set.

(2)   Extend c_1 to a cut $c \in C$, i.e., $c_1 \subseteq c$.

(3)   By K-density we get: $l_0 \cap c = \{a\}$ for some $a \in X$.

(4)   For this $a \in X$ we have the following possibilities:

   (a) $a \prec x_0$; (b) $a \succ y_0$; (c) $a \in [x_0, y_0]$.

   We shall prove that from all of these possibilities, a contradiction follows.

   First we prove the following:

Claim A: $X_1 \subseteq \downarrow c_1$ and $Y_1 \subseteq \uparrow c_1$.

Proof: We shall prove only the first inclusion. The proof of $Y_1 \subseteq \uparrow c_1$ is similar. Let $x \in X_1$ and take any $y \in Y_1$. We know that x li y and $x \neq y$. If $y \prec x$ then $\downarrow y \subseteq \downarrow x$ and also $l_1 \cap \downarrow y \subseteq l_1 \cap \downarrow x$. From this it follows that $|l_1 \cap \downarrow y| \leq |l_1 \cap \downarrow x| \in \mathbf{N}^+$ which is a contradiction since $y \in Y_1 = l_1 \setminus X_1$. So, we must have $x \prec y$. (Note that this property is true for all $x \in X_1$ and $y \in Y_1$.) Since $(X, \prec)$ is combinatorial there exists a finite sequence

$$x = x_1 \prec x_2 \prec \ \ldots\ \prec x_n = y.$$

Since $x = x_1 \in \downarrow X_1$ and $y = x_n \notin \downarrow X_1$, there exists $m \in \mathbf{N}^+$, $1 \leq m < n$, such that $x_m \in \downarrow X_1$ and $x_{m+1} \notin \downarrow X_1$. By the definition of c_1, we have $x_{m+1} \in c_1$. But we know that $x \prec x_{m+1}$. Then $x \in \downarrow c_1$, i.e., in all, $X_1 \subseteq \downarrow c_1$.   □ *Claim A*

(1)   We prove now that c_1 is a co-set. First of all, $c_1 \neq \emptyset$ because of $\emptyset \neq X_1 \subseteq \downarrow c_1$.

We conduct an indirect proof. Assume that there exist $u, v \in c_1$ with $u \prec v$. By the definition of c_1, $\exists w \in \downarrow X_1$ such that $w \prec v$. Since $Max(X_1) = \emptyset$, $\exists z \in X_1$ such that $w \prec z$ (see Figure 2.11).

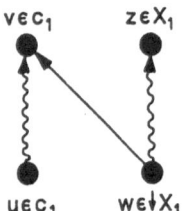

$v \in c_1$ $z \in X_1$

$u \in c_1$ $w \in \downarrow X_1$

Fig. 2.11. Illustration of the proof of (1)

The idea now is to show that $u \ co \ w$, $v \ co \ z$ and $u \ co \ z$. Then since $(X, \prec)$ is N-dense, there must be an element strictly between w and v, which would be a contradiction since $w \prec\hspace{-1.3mm}\cdot\ v$.

$u \ co \ w$: $u \prec w \in\downarrow X_1$ implies $u \in\downarrow X_1$ which is a contradiction since $u \in c_1$ (by the definition of c_1). $w \prec u$ implies $w \prec u \prec v$, a contradiction with $w \prec\hspace{-1.3mm}\cdot\ v$. I.e., we must have $u \ co \ w$.

$v \ co \ z$: $z \prec v$ implies $w \prec z \prec v$, a contradiction with $w \prec\hspace{-1.3mm}\cdot\ v$. $v \prec z \in X_1$ implies $v \in\downarrow X_1$, a contradiction since $v \in c_1$ (by the definition of c_1). We must have $v \ co \ z$.

$u \ co \ z$: $u \prec z \in X_1$ implies $u \in\downarrow X_1$, again a contradiction since $u \in c_1$. $z \prec u$ implies $z \prec v$, a contradiction since we have just proved that $v \ co \ z$. Hence $u \ co \ z$.

This finishes the proof of (1), i.e., c_1 is a co-set.

(2)   Let $c \in C$ be any cut such that $c \supseteq c_1$.

(3)   By K-density, $c \cap l_0 = \{a\}$ for some $a \in X$.

(4)   We have to consider three cases:

$a \prec x_0$:   We know that $x_0 \in\downarrow c_1$ by Claim A. Then $\exists y \in c_1$ such that $x_0 \preceq y$. This implies $a \prec x_0 \preceq y \in c_1$, a contradiction since $a, y \in c$.

$a \succ y_0$: $y_0 \in\uparrow c_1$, again by Claim A. So, $\exists y \in c_1 \subseteq c$ such that $y \preceq y_0$. This implies $y \preceq y_0 \prec a$, a contradiction since $y, a \in c$.

$a \in [x_0, y_0]$:   In this case, $a \in l_1 = l_0 \cap [x_0, y_0]$. But $l_1 = X_1 \cup Y_1$ and $X_1 \subseteq\downarrow c_1 \wedge Y_1 \subseteq\uparrow c_1$ (Claim A) implies $a \in l_1 \subseteq\downarrow c_1 \cup \uparrow c_1$, i.e., $a \in\downarrow c_1 \cup \uparrow c_1$. From this it follows immediately that $a \in c_1$.

Claim B: $X_1 \subseteq \downarrow a$.

Proof: Let $x \in X_1$. Then $x \ li \ a$ since $x, a \in l_1$. By Claim A we have: $X_1 \subseteq\downarrow c_1$. In particular, $x \in\downarrow c_1$, i.e., $x \preceq y$ for some $y \in c_1$. If $a \prec x$ then $a \prec y$, a contradiction since $a, y \in c_1$. It follows that $x \preceq a$, i.e., $x \in\downarrow a$.   □ *Claim B*

Finally, $a \in c_1$ implies $\exists y \in\downarrow X_1$ such that $y \prec\hspace{-1.3mm}\cdot\ a$. $y \in\downarrow X_1 \Rightarrow \exists z \in X_1$ such that $y \preceq z$. We know that $z \prec a$ since $z \in X_1$, $a \in c_1$ and Claim B. If $y \prec z$ then $y \prec z \prec a$, a contradiction since $y \prec\hspace{-1.3mm}\cdot\ a$. If $y = z \in X_1$ then $\exists z_1 \in X_1$ such that $y \prec z_1$ since $Max(X_1) = \emptyset$. Again $z_1 \prec a$ and $y \prec z_1 \prec a$, a contradiction.   □ *Case 1*

Case 2: $Min(Y_1) = \emptyset$. In this case we define:

$$c_2 = \{x \in X \mid \exists y \in\uparrow Y_1 : x \prec\hspace{-1.3mm}\cdot\ y\} \setminus (\uparrow Y_1),$$

and we proceed analogously to Case 1.   □ *Case 2*

□ 2.3.7

Proposition 2.3.8

If $(X, \prec)$ is weakly discrete and has the line crossing property then none of the two posets shown in Figure 2.10(i,ii) is embeddable into $(X, \prec)$.

Proof: (Indirect.) Assume that the poset of Figure 2.10(i) is embeddable into $(X, \prec)$. (If the poset shown in Figure 2.10(ii) is embeddable then the proof is similar.) Then there exist two subsets of X,

$$X_1 = \{x_1, x_2, x_3, \ldots\} \subseteq X$$
$$\text{and} \quad Y_1 = \{y_1, y_2, y_3, \ldots\} \subseteq X$$

with the following properties:

(a) $|X_1| = |Y_1| \notin \mathbf{N}$.

(b)   If $i < j$ then $x_i \prec x_j$.

(c) $\forall i, j \in \mathbf{N}^+ : y_i$ co y_j and ($y_i \neq y_j$ iff $i \neq j$).

(d)   If $i \leq j$ then $x_i \prec y_j$.

(e)   If $i > j$ then x_i co y_j and $x_i \neq y_j$.

(See Figure 2.12.)

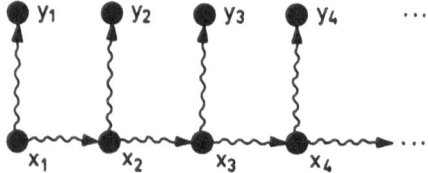

Fig. 2.12. Illustration of the proof of Proposition 2.3.8

X_1 is a li-set by (b) and Y_1 is a co-set by (c). Let $l \in L$ be a line such that $X_1 \subseteq l$ and let $c \in C$ be a cut such that $Y_1 \subseteq c$. Since $(X, \prec)$ has the line crossing property, we have $l \cap \uparrow c \neq \emptyset$ ($l \cap \downarrow c \neq \emptyset$ is used in the dual proof).

Let $x \in l \cap \uparrow c$.

Claim C: $x_i \prec x$ for all $i \in \mathbf{N}^+$.

Proof: $x \in \uparrow c \Rightarrow \exists y \in c : y \preceq x$. x li x_i since $x, x_i \in l$. If $x \preceq x_i$ then $x \preceq x_i \prec y_i \in Y_1 \subseteq c$. But then $y \preceq x \preceq x_i \prec y_i$, a contradiction since $y, y_i \in c$. Hence we must have $x_i \prec x$ for all $i \in \mathbf{N}^+$.   □ *Claim C*

Then we have $x_i \in l \cap [x_1, x]$ for all $i \in \mathbf{N}^+$, i.e., $X_1 \subseteq l \cap [x_1, x]$. But $|X_1| \notin \mathbf{N}$; then $|l \cap [x_1, x]| \notin \mathbf{N}$, which is a contradiction with the fact that $(X, \prec)$ is weakly discrete.   □ 2.3.8

Proposition 2.3.9
If none of the two posets shown in Figure 2.10(i,ii) is embeddable into $(X, \prec)$ then $(X, \prec)$ is K-dense.

Proof: (Indirect.) Assume that $(X, \prec)$ is not K-dense.
Then $\exists l \in L \, \exists c \in C: l \cap c = \emptyset$. Only three cases are possible:

Either (1) $l \subseteq \downarrow c$ or (2) $l \subseteq \uparrow c$ or (3) $l \cap \downarrow c \neq \emptyset \wedge l \cap \uparrow c \neq \emptyset$.

Case 1: $l \subseteq \downarrow c$.
Firstly, we claim that $Max(l) = \emptyset$. If not, let $x_0 \in Max(l)$. $x \preceq x_0$ for all $x \in l$. On the other hand, $x_0 \prec y_0$ for some $y_0 \in c$ since $l \subseteq \downarrow c$; i.e., $x \preceq x_0 \prec y_0$ for all $x \in l$, a contradiction to the fact that l is a line. This proves the claim that $Max(l) = \emptyset$.

Let $x_1 \in l$. Then $\exists y_1 \in c$ such that $x_1 \prec y_1$. Because $Max(l) = \emptyset$ there exists $x \in l$ such that $x \succ x_1$. If $x \prec y_1$ for all $x \in l$ then $y_1 \in l$, a contradiction. This means that we can choose $x_2 \in l$ and $y_2 \in c$ such that:

$$x_1 \prec x_2 \prec y_2 \text{ and } y_1 \text{ co } x_2 \text{ and } y_1 \neq y_2.$$

(See Figure 2.13.)

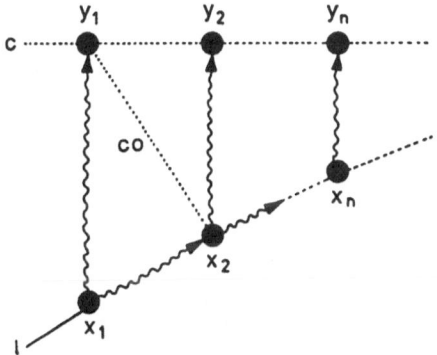

Fig. 2.13. Illustration of Case 1

We proceed now by induction. Assume that there exist $x_n \in l$, $y_n \in c$ such that

(j) $x_1 \prec x_2 \prec \ldots \prec x_n \prec y_n$;

(jj) $y_i \text{ co } x_n$ for all $i = 1, 2, \ldots, n-1$;

(jjj) $y_i \neq y_j$ for $i \neq j$.

Again because of $Max(l) = \emptyset$ there exists $x \in l: x \succ x_n$. If $x \prec y_n$ for all $x \in l$ then $y_n \in l$, a contradiction. On the other hand, if $x \succ x_n$ then x co y_i for $1 \leq i < n$ because if $x \prec y_i$ for $1 \leq i < n$ then $x_n \prec y_i$, a contradiction with (jj). All of this means that there exists $x_{n+1} \in l$ and $y_{n+1} \in c$ such that

$$x_1 \prec x_2 \prec \ldots \prec x_n \prec x_{n+1} \prec y_{n+1}$$
$$\wedge \; y_i \text{ co } x_{n+1} \; (i = 1, 2, \ldots, n) \; \wedge \; y_i \neq y_j \text{ for } i \neq j.$$

In fact with the help of the induction we have proved that there exist two sets $X_1 = \{x_1, x_2, \ldots\} \subseteq l$ and $Y_1 = \{y_1, y_2, \ldots\} \subseteq c$ with the following properties:

(a) $|X_1| = |Y_1| \notin \mathbf{N}$;

(b)   If $i < j$ then $x_i \prec x_j$;

(c) $\forall i, j \in \mathbf{N}^+: y_i$ co y_j and $(y_i \neq y_j$ iff $i \neq j)$;

(d)   If $i \leq j$ then $x_i \prec y_j$;

(e)   If $i > j$ then x_i co y_j and $x_i \neq y_j$.

The poset shown in Figure 2.10(i) could clearly be embedded into the poset $(X', \prec |_{X' \times X'})$ where $X' = X_1 \cup Y_1$, yielding a contradiction with the hypothesis.   □ *Case 1*

Case 2: $l \subseteq \uparrow c$.
In this case we have, by similarity with Case 1, that $Min(l) = \emptyset$. Proceeding with a similar construction to the one applied in Case 1 we get two sets $X_2 = \{x_1, x_2, \ldots\} \subseteq l$ and $Y_2 = \{y_1, y_2, \ldots\} \subseteq c$ with the following properties:

(a) $|X_2| = |Y_2| \notin \mathbf{N}$;

(b)   If $i < j$ then $x_i \succ x_j$;

(c) $\forall i, j \in \mathbf{N}^+: y_i$ co y_j and $(y_i \neq y_j$ iff $i \neq j)$;

(d)   If $i \leq j$ then $x_i \succ y_j$;

(e)   If $i > j$ then x_i co y_j and $x_i \neq y_j$.

In this case, the poset shown in Figure 2.10(ii) could clearly be embedded into the poset $(X'', \prec |_{X'' \times X''})$ where $X'' = X_2 \cup Y_2$, which is again a contradiction with the hypothesis.   □ *Case 2*

Case 3: $l \cap \downarrow c \neq \emptyset \wedge l \cap \uparrow c \neq \emptyset$.
In this case there exist two subsets l_1 and l_2 of l such that $\emptyset \neq l_1 \subseteq l$ and $\emptyset \neq l_2 \subseteq l$ and $l_1 \cup l_2 = l$ and $l_1 \subseteq \downarrow c$ and $l_2 \subseteq \uparrow c$. Firstly we prove that if $Max(l_1) \neq \emptyset$ and $Min(l_2) \neq \emptyset$ then we arrive at a contradiction.

Let $x' \in Max(l_1)$ and $x'' \in Min(l_2)$ (see Figure 2.14).

Since $x', x'' \in l$ it follows clearly that $x' \prec \cdot x''$. $x' \in l_1 \subseteq \downarrow c$ implies $\exists y' \in c: x' \prec y'$. $x'' \in l_2 \subseteq \uparrow c$ implies $\exists y'' \in c: y'' \prec x''$.

Now, $y' \prec x''$ is a contradiction with $x' \prec x''$; $y' \succ x''$ implies $y' \succ x'' \succ y''$, a contradiction since $y', y'' \in c$; i.e., y' co x''.

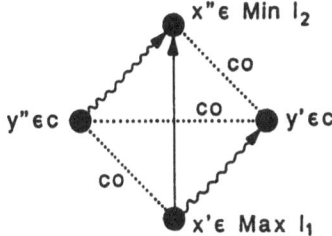

Fig. 2.14. Illustration of the proof that $\neg(Max(l_1) \neq \emptyset \wedge Min(l_2) \neq \emptyset)$

In the same way, $x' \prec y''$ is a contradiction with $x' \prec\!\!\cdot\, x''$; $x' \succ y''$ implies $y'' \prec y'$, a contradiction since $y', y'' \in c$; i.e., $x'\ co\ y''$.

Also, $y' \neq y''$ because of $x' \prec\!\!\cdot\, x''$. But then since $(X, \prec)$ is N-dense there exists $x' \prec x \prec x''$ which is again a contradiction with $x' \prec\!\!\cdot\, x''$. This shows that either $Max(l_1) = \emptyset$ or $Min(l_2) = \emptyset$ or both.

Assume that $Max(l_1) = \emptyset$. (The other case is exactly dual.) If $x \in l_1$ and $y \in l_2$ then we must have $x \prec y$. We prove now the following claim:

Claim D: If $x \in l_1$ and $y \in l_2$ then there exists $z \in c$ such that $x \prec z \prec y$.

Proof: Since $(X, \prec)$ is combinatorial there exists a finite set $\{x_1, x_2, \ldots, x_n\}$ such that $x = x_1 \prec\!\!\cdot\, x_2 \prec\!\!\cdot\, \ldots \prec\!\!\cdot\, x_n = y$. Since $x \in l_1 \subseteq\,\downarrow\!c$ and $y \in l_2 \subseteq\,\uparrow\!c$, there exists $m \in \mathbf{N}^+$, $1 \leq m < n$, such that $x_m \in\,\downarrow\!c$ and $x_{m+1} \in\,\uparrow\!c$ (see Figure 2.15).

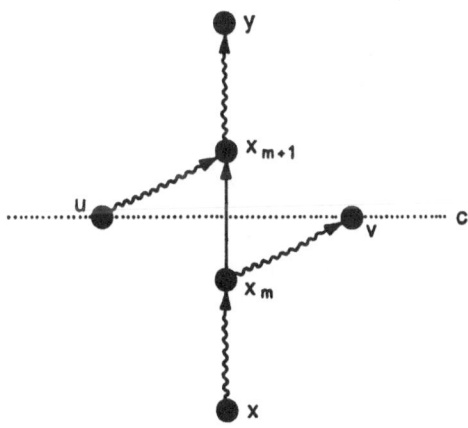

Fig. 2.15. Illustration of the proof of Claim D

If $x_m \in c$ or $x_{m+1} \in c$ then the claim is proved. If $x_m \notin c$ and $x_{m+1} \notin c$ then $\exists u, v \in c$ such that $u \prec x_{m+1}$ and $v \succ x_m$. $u = v$ contradicts $x_m \prec x_{m+1}$;

i.e., $u \neq v$. Clearly, v co x_{m+1} and u co x_m.

So, by N-density, $\exists z \in X : x_m \prec z \prec x_{m+1}$, a contradiction with $x_m \not\prec x_{m+1}$. I.e., either $x_m \in c$ or $x_{m+1} \in c$, and the claim is true. $\square$ *Claim D*

Now assume that for all $x \in l_1$ there exists a triangle structure as shown in Figure 2.16.

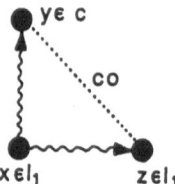

Fig. 2.16. Illustration of property $\boxed{A}$

Formally:

$\boxed{A}$ $\forall x \in l_1 \; \exists y \in c \; \exists z \in l_1 : x \prec y \land y$ co z.

Clearly, $x \neq z$ and $x \prec z$. Let $x_0 \in l_1$. By $\boxed{A}$, $\exists y_0 \in c \; \exists x_1 \in l_1$ such that $x_0 \prec y_0$ co x_1. Again, $x_1 \in l_1$ implies $\exists y_1 \in c \; \exists x_2 \in l_1$ such that $x_1 \prec y_1$ co x_2. Note that, clearly, $y_0 \neq y_1$. Continuing with this construction we get two subset of X,

$$X_1 = \{x_0, x_1, \ldots\} \text{ and } Y_1 = \{y_0, y_1, \ldots\}$$

which satisfy the properties (a)–(e) listed in the proof of Proposition 2.3.8 above (see Figure 2.17).

This implies that the poset shown in Figure 2.10(i) can be embedded into the poset $(X', \prec |_{X' \times X'})$ where $X' = X_1 \cup Y_1$, a contradiction with the hypothesis.

Finally, assume the negation of $\boxed{A}$, i.e., assume there exists $x \in l_1$ for which it is not possible to construct a triangle structure. Formally:

$\boxed{B} = \neg \boxed{A}$ $\exists x \in l_1 \; \forall y \in c \; \forall z \in l_1 : \neg(x \prec y) \lor \neg(y$ co $z)$.

An equivalent form of $\boxed{B}$ is the following:

$\boxed{B}$ $\exists x \in l_1 \; \forall y \in c \; \forall z \in l_1 : x \prec y \Rightarrow y$ li z.

Let $x_0 \in l_1$ be such that $\boxed{B}$ is true for $x = x_0$ (see Figure 2.18).

$x_0 \in l_1 \subseteq \downarrow c$ implies $\exists y_0 \in c$ such that $x_0 \prec y_0$. By $\boxed{B}$, $\forall x \in l_1 : x$ li y_0. Since $l_1 \subseteq \downarrow c$, it is clear that $x \prec y_0$. $y_0 \notin l$ implies $\exists z_0 \in l$ such that y_0 co z_0.

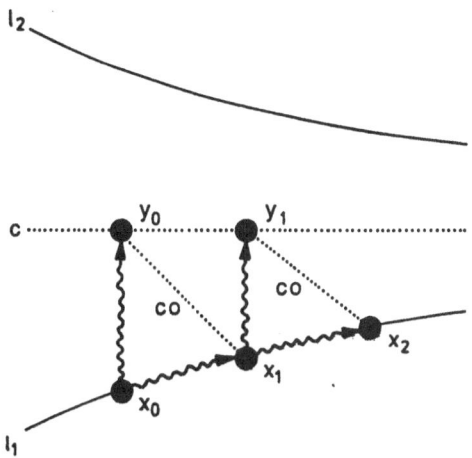

.**Fig. 2.17.** Illustration of the proof of Case 3 if $\boxed{A}$ holds

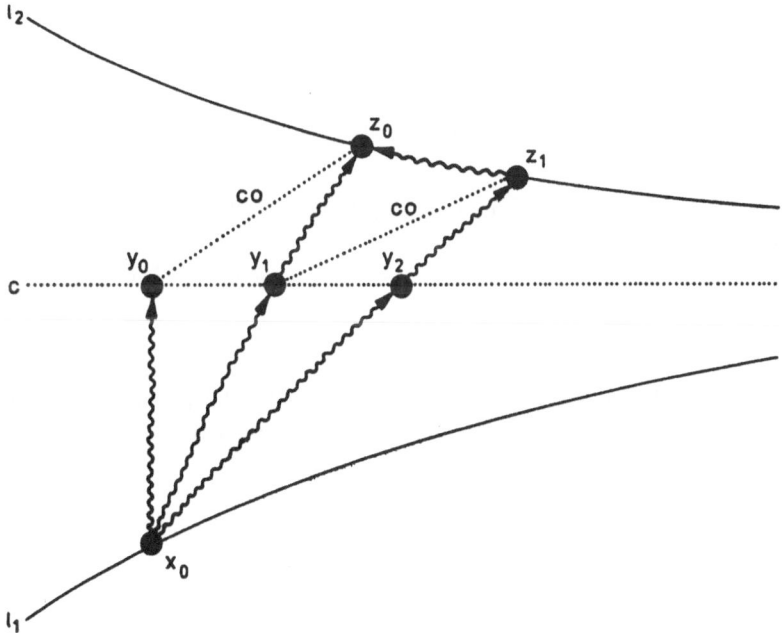

Fig. 2.18. Illustration of the proof of Case 3 if $\boxed{B}$ holds.

By $\boxed{B}$, $z_0 \in l_2$. By Claim D, $\exists y_1 \in c$ such that $x_0 \prec y_1 \prec z_0$. Clearly, $y_0 \neq y_1$. Furthermore $\forall x \in l_1: x \prec y_1$ since $\boxed{B}$ is true. There is an element $z_1 \in l_2$ such that z_1 co y_1. Again by Claim D, $\exists y_2 \in c: x_0 \prec y_2 \prec z_1$. We can continue with this construction. If it finishes at one point (i.e., if $Min(l_2) = \{z\}$) then $\exists y \in c: x_0 \prec y \prec z \in Min(l_2)$. This means that y li x for all $x \in l$, i.e., $y \in l$, a contradiction since $l \cap c = \emptyset$.

If the construction does not finish then there exist two subsets of X,

$$Z = \{z_0, z_1, \ldots\} \text{ and } Y = \{y_0, y_1, \ldots\}$$

which again satisfy the necessary properties such that the poset of Figure 2.10(ii) can be embedded into the poset $(X'', \prec |_{X'' \times X''})$ where $X'' = Y \cup Z$. Again this contradicts the hypothesis, and the proof is finally done.   □ *Case 3*
□ 2.3.9

Remark 2.3.10 [Stronger results]
Combinatorialness and N-density are not always necessary for the various statements of this theorem (see Exercise 19 at the end of this chapter). However, we shall be content with the theorem as it stands because it is already rather general in the form given here; later, we shall have reasons to deal with combinatorialness and N-density as general presumptions, so that the preconditions of the theorem are then satisfied.   □ 2.3.10

We state a corollary which gives sufficient conditions for K-density:

Corollary 2.3.11 [The special case of finite cuts and finite lines]
Let $(X, \prec)$ be a combinatorial and N-dense poset.
If $(X, \prec)$ has only finite lines or if $(X, \prec)$ has only finite cuts then $(X, \prec)$ is K-dense.

Proof: From Proposition 2.3.9 and the fact that the posets shown in Figure 2.10(i,ii) have both an infinite line and an infinite cut.   □ 2.3.11

There is a partial converse of this corollary which allows the finiteness of all cuts to be deduced from K-density (together with some other assumptions). Since we will need this result later, we will state it here.

Theorem 2.3.12 [Deduction of cut-finiteness from K-density]
Let $(X, \prec)$ be a poset which is K-dense, degree-finite, combinatorial, and has a finite cut. Then all cuts of $(X, \prec)$ are finite.

Proof: Let c_0 be a finite cut that exists by hypothesis and let c be any arbitrary cut. For $x \in c_0$ consider the set

$$Y_x = \{y \in c \mid x \, li \, y\}.$$

We have either $Y_x = (c \cap \uparrow x)$ or $Y_x = (c \cap \downarrow x)$ because c is a co-set, and furthermore, $Y_x \neq \emptyset$ because c is co-maximal.

Suppose first that $Y_x = (c \cap \uparrow x)$; we wish to prove that Y_x is a finite set. To this end we assume Y_x to be infinite and derive a contradiction. We may construct a li-set $\{x_0, x_1, x_2, \ldots\}$ inductively as follows. Put $x_0 = x$. By induction hypothesis, there are infinitely many $y \in Y_x$ such that $x_i \preceq y$. By degree-finiteness, there is some $x_{i+1} \in x_i^{\bullet}$ such that, again, $x_{i+1} \preceq y$ for infinitely many $y \in Y_x$ (this argument recurs several times in this book, for instance in the proof of Theorem 2.2.10(iii), but also in later chapters). There is a line $l \supseteq \{x_0, x_1, x_2, \ldots\}$. By K-density, $\exists z \in X : z \in c \cap l$. But then we have $|[x, z] \cap l| \notin \mathbf{N}$, contradicting Proposition 2.3.7. Hence we have arrived at a contradiction, which shows that Y_x is finite.

If $Y_x = c \cap \downarrow x$ then the proof that Y_x is finite proceeds analogously. But because c_0 is a cut, we have:

$$c = \bigcup_{x \in c_0} Y_x.$$

Since both c_0 and Y_x are finite, so is c. $\square$ 2.3.12

2.4 D-continuity

The property which will be studied in this section is based on an analogy between the lines of a poset (and their interpretation as sequential subprocesses) and the notion of a world line in Physics. A physical happening is often described as a mesh of the world lines of interacting particles in the same way as a partially ordered set can be imagined to be a mesh of its lines. In physical modelling, the world line of an individual particle is described by a continuous curve with properties akin to those of the line of the reals.

In particular, the real line satisfies a special density property which distinguishes it both from the integers and from the rationals: there are no jumps as in $\mathbf{Z}$, say between the set $\{x \in \mathbf{Z} \mid x \leq 2\}$ and the set $\{x \in \mathbf{Z} \mid x \geq 3\}$; also, there are no gaps as in the set $\mathbf{Q}$ of rationals, say between the set $\{x \in \mathbf{Q} \mid x \leq 0 \vee x^2 < 2\}$ and the set $\{x \in \mathbf{Q} \mid x > 0 \wedge x^2 > 2\}$ (i.e., at the place of $\sqrt{2}$). The fact that the reals are 'continuous' (complete) in this sense is often known as the Dedekind completeness property of the reals, following the well known Dedekind cut construction of the real numbers. In this section, we will define a property called D-continuity which aims at generalising the Dedekind completeness property of the reals to arbitrary (in particular, to discrete) posets.

To introduce this property, let us first recall the definition of a Dedekind cut. The set of reals, $\mathbf{R}$, with the usual ordering $<$ is a linear order (i.e., $co = id|_{\mathbf{R}}$); it has one line, $\mathbf{R}$ itself, and each real number $r \in \mathbf{R}$ forms a singleton cut $\{r\}$. Any partitioning of $\mathbf{R}$ into two disjoint sets $\mathbf{R}_1$ and $\mathbf{R}_2 = \mathbf{R} \setminus \mathbf{R}_1$ such that no number in $\mathbf{R}_2$ is smaller than any number in $\mathbf{R}_1$ is called a Dedekind cut. We transport this idea to posets in the following way:

Definition 2.4.1 [Dedekind cuts]

(i) For $A \subseteq X$ we define $\overline{A} = X \setminus A$.

(ii) A pair $(A, \overline{A})$ is a Dedekind cut (D-cut for short) *iff*
$A \neq \emptyset \neq \overline{A}$ and $\forall x \in A \ \forall y \in \overline{A}: \neg(y \preceq x)$.
The set of all D-cuts of $(X, \prec)$ will be denoted by $D = D(X, \prec)$.   □ 2.4.1

According to this definition, $(A, \overline{A})$ is a non-trivial partitioning of X which in the sense of Definition 2.4.1(ii) respects the $\preceq$ relation on X. It can be seen that if $(A, \overline{A})$ is a D-cut then $A = \downarrow A$ and $\overline{A} = \uparrow \overline{A}$. Conversely, if $A \subseteq X$ such that $\emptyset \neq A \neq X$ and $\downarrow A = A$ then the pair $(A, \overline{A})$ satisfies the definition of a D-cut. Hence we are justified in abusing the terminology slightly and saying that 'A is a D-cut' rather than '$(A, \overline{A})$ is a D-cut'.
We also need the following definition:

Definition 2.4.2 [The set $M(A)$]
For $A \subseteq X$, $M(A) = Max(A) \cup Min(\overline{A})$.   □ 2.4.2

As an example, consider the poset shown in Figure 2.19 with the partitioning $(A, \overline{A})$. $(A, \overline{A})$ is clearly a D-cut and we have $Max(A) = \{x, y\}$, $Min(\overline{A}) = \{u, v\}$, $M(A) = \{x, y, u, v\}$.
An immediate generalisation of the Dedekind construction would be to call a poset D-continuous if $|M(A) \cap l| = 1$ for every D-cut A and every line l, because A and $\overline{A}$ correspond to $\mathbf{R}_1$ and $\mathbf{R}_2$ above and $M(A)$ is the set of elements of X that 'separates' them:

Definition 2.4.3 [An attempted definition of D-continuity]
$(X, \prec)$ is D'-continuous *iff* $\forall A \in D \ \forall l \in L: |M(A) \cap l| = 1$.   □ 2.4.3

Indeed, as can easily be seen, $(\mathbf{R}, <)$ (but neither $(\mathbf{Q}, <)$ nor $(\mathbf{Z}, <)$) is D'-continuous. However, we have the following.

Lemma 2.4.4 [D'-continuity implies density]
If $(X, \prec)$ is D'-continuous then $(X, \prec)$ is dense.

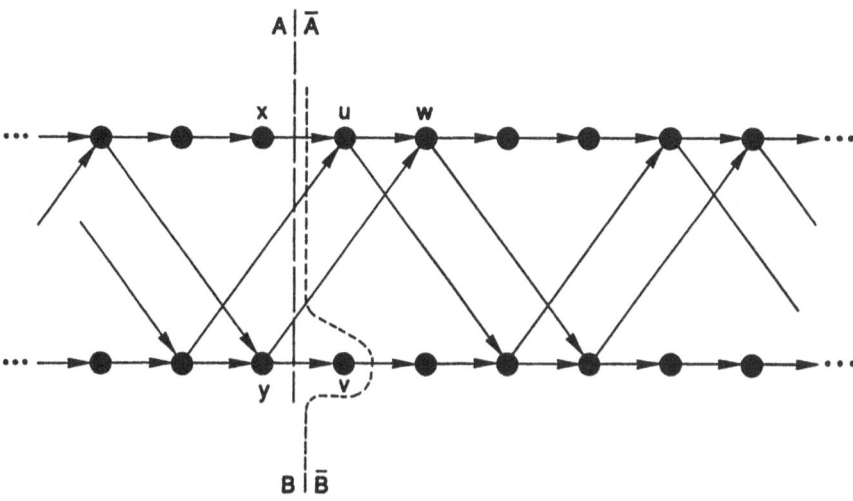

Fig. 2.19. An example illustrating D-cuts

Proof: We have to prove that $\prec = \emptyset$; we do an indirect proof.
Assume that $\prec \neq \emptyset$, then $\exists x, y \in X : x \prec y$.
Define $A = \downarrow y \setminus \{y\}$ and $\overline{A} = X \setminus A$.
Clearly, $(A, \overline{A})$ is a D-cut.
Furthermore: $x \in Max(A)$ since $x \prec y$ and $A = \downarrow y \setminus \{y\}$; also: $y \in Min(\overline{A})$ by
the construction of A.
Let $l \in L$ be such that $x, y \in l$.
Then $|M(A) \cap l| = 2$, a contradiction with the fact that $(X, \prec)$ is
D'-continuous. $\square$ 2.4.4

  Because of this, D'-continuity is not suited as a generalisation of the Dedekind
construction to discrete posets (i.e., posets that satisfy any of the properties
introduced in Section 2.2). One possibility is to narrow down the set $M(A)$:

Definition 2.4.5 [The sets $Obmax$, $Obmin$, $c(A)$ and D-continuity]

(i)   For $A \in D(X, \prec)$,

$Obmax(A) =$
$\{x \in Max(A) \mid \forall A' \in D \; \forall l \in L : x \in Max(A' \cap l) \Rightarrow x \in Max(A')\};$
$Obmin(\overline{A}) =$
$\{x \in Min(\overline{A}) \mid \forall A' \in D \; \forall l \in L : x \in Min(\overline{A'} \cap l) \Rightarrow x \in Min(\overline{A'})\};$
$c(A) = Obmax(A) \cup Obmin(\overline{A}).$

(ii) $(X, \prec)$ is D-continuous *iff* $\forall A \in D \; \forall l \in L : |c(A) \cap l| = 1.$ $\square$ 2.4.5

The idea behind the *Obmax* definition is that of an element being objectively maximal in the sense that 'all lines agree on its maximality'; and similarly for *Obmin*. In the example shown in Figure 2.19 (with the partitioning shown there), we have $Obmax(A) = \{x\}$, $Obmin(\overline{A}) = \{v\}$ and $c(A) = \{x, v\}$.

In order to see $y \notin Obmax(A)$, it suffices to choose a D-cut $(B, \overline{B})$ as indicated in the figure; y is in $Max(l' \cap B)$ where l' is the line through y and w, but it is not in $Max(l'' \cap B)$ where l'' is the line through y and v, and hence $y \notin Max(B)$. On the other hand, x is a maximal point of A, independently of the cut in its neighbourhood, and hence $x \in Obmax(A)$.

The following proposition gives a characterisation of the sets *Obmax* and *Obmin*.

Proposition 2.4.6 [Characterisation of *Obmax* and *Obmin*]
Let $A \in D(X, \prec)$ and $x \in Max(A)$, $y \in Min(\overline{A})$.

(i) $x \notin Obmax(A) \iff \exists z \in X\ \exists l \in L\colon x \prec z \wedge l \cap [x, z] = \{x\}$;

(ii) $y \notin Obmin(\overline{A}) \iff \exists z \in X\ \exists l \in L\colon z \prec y \wedge l \cap [z, y] = \{y\}$.

Proof:

(i)   ($\Rightarrow$)   If $x \notin Obmax(A)$ then for some $A' \in D$ and some $l \in L$ we have $x \in Max(A' \cap l)$ and $x \notin Max(A')$. $x \notin Max(A')$ implies that $\exists z \in A'$ such that $x \prec z$. We claim that $l \cap [x, z] = \{x\}$. Assume that there exists $w \neq x$ such that $w \in l \cap [x, z]$. Since $w \preceq z$ and $z \in A'$ it follows that $w \in A'$. But then we have $w \in A' \cap l$ and $w \succ x$ which is a contradiction with the fact that $x \in Max(A' \cap l)$.

($\Leftarrow$)   Let $z \in X$ and $l \in L$ such that $x \prec z$ and $l \cap [x, z] = \{x\}$. Define $A' = \downarrow z$. From the definition it follows that A' is a D-cut.

Claim: $x \in Max(A' \cap l)$.

Proof: Clearly $x \in A' \cap l$. If $x \notin Max(A' \cap l)$ then $\exists w \in A' \cap l$ such that $x \prec w$. $w \preceq z$ since $w \in A' = \downarrow z$. I.e., $w \in [x, z]$. It follows that $w \neq x$ and $w \in l \cap [x, z]$, which is a contradiction with the hypothesis.   □ *Claim*

Furthermore, $x \notin Max(A')$ since $z \in A'$ and $x \prec z$. Hence for the D-cut A' we have that $x \in Max(A' \cap l)$ but $x \notin Max(A')$. This means, by definition, that $x \notin Obmax(A)$.

(ii)   Similar proof.   □ 2.4.6

For combinatorial posets, the characterisation is much simpler. We get the following corollary:

Corollary 2.4.7

Let $(X, \prec)$ be a combinatorial poset. Then:

(i) $Obmax(A) = \{x \in Max(A) \mid |x^\bullet| \leq 1\};$

(ii) $Obmin(\overline{A}) = \{x \in Min(\overline{A}) \mid |^\bullet x| \leq 1\}.$

Proof: Follows easily from Proposition 2.4.6.   □ 2.4.7

This corollary allows to recognise easily which elements belong to $c(A)$. In fact, if $x \in Max(A)$ then $x \in Obmax(A)$ if x does not have two or more immediate successors, and vice versa for $Obmin$.

In order to obtain a characterisation of D-continuity, it is meaningful to distinguish the two cases $|c(A) \cap l| \neq 0$ and $|c(A) \cap l| \neq 2$. For linearly ordered sets, these two cases reduce to $|M(A)| \neq 0$ and $|M(A)| \neq 2$, respectively. For example, in the poset of the rationals $(\mathbf{Q}, <)$, we have

$$|M(\{x \in \mathbf{Q} \mid x \leq 0 \vee x^2 < 2\})| = 0,$$

This may be called a 'gap'. On the other hand, in $(\mathbf{Z}, <)$ we have

$$|M(\{x \in \mathbf{Z} \mid x \leq 2\})| = |\{2, 3\}| = 2,$$

This may be called a 'jump', justifying the following definition.

Definition 2.4.8 [Gap-freeness and jump-freeness]

(i) $(X, \prec)$ is free of gaps (gap-free) *iff* $\forall A \in D \; \forall l \in L: |c(A) \cap l| \neq 0.$

(ii) $(X, \prec)$ is free of jumps (jump-free) *iff* $\forall A \in D \; \forall l \in L: |c(A) \cap l| \neq 2.$
                    □ 2.4.8

As an example, consider Figure 2.20. The poset shown in Figure 2.20(i) is free of gaps but not free of jumps, since for the D-cut $(A, \overline{A})$ with $A = \{x_0, x_1, x_2\}$ and $\overline{A} = \{x_3, x_4, x_5\}$ we have $c(A) = \{x_2, x_3\}$, so $|c(A) \cap l| = 2$ for any line containing both x_2 and x_3. On the other hand, the poset shown in Figure 2.20(ii) is free of jumps but not free of gaps; the demonstration of this fact is left as an exercise for the reader.

Clearly, $(X, \prec)$ is D-continuous iff it is both free of gaps and free of jumps, since, as is easy to see, the case $|c(A) \cap l| > 2$ can never arise. We shall characterise the properties defined in Definition 2.4.8 separately. If $|c(A) \cap l| = 2$ then there must be elements $x, y \in c(A)$ such that not only $x \not\prec y$ but also no line through x can bypass y and no line through y can bypass x. By requiring the contrary, the case $|c(A) \cap l| = 2$ can be prohibited. For this purpose we define the property of non-single degree; as it will turn out, this property eliminates jumps in general posets in the same way as density eliminates jumps in totally ordered sets. The name of the property will be explained in Section 2.5.

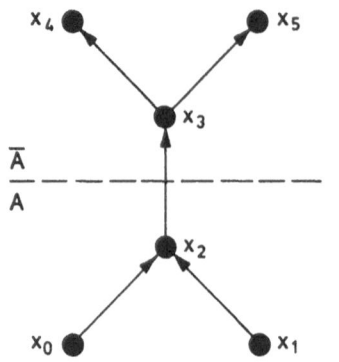
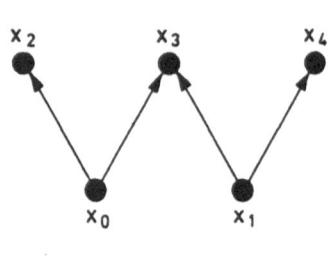

(i) Free of gaps but not free of jumps   (ii) Free of jumps but not free of gaps

Fig. 2.20. Illustration of jump-freeness and gap-freeness

Definition 2.4.9 [The non-single degree property]
$(X, \prec)$ is of non-single degree *iff*
$$\forall x, y \in X : x \prec y \Rightarrow \exists z \in X, x \neq z \neq y : (x \prec z \text{ co } y) \lor (x \text{ co } z \prec y). \qquad \Box \ 2.4.9$$

Theorem 2.4.10 [Characterisation of jump-freeness]
$(X, \prec)$ *has no jumps iff* $(X, \prec)$ *is of non-single degree.*

Proof:

$(\Rightarrow)$ Let $x, y \in X$ be such that $x \prec y$.
We have to prove that $\exists z \in X$, $x \neq z \neq y$ such that $x \prec z \text{ co } y$ or $x \text{ co } z \prec y$.
     Define $A = \downarrow x$ and $\overline{A} = X \setminus A$. Clearly, $(A, \overline{A})$ is a D-cut and $x \in Max(A)$.
If $y \notin Min(\overline{A})$ then $\exists z \in \overline{A}$ such that $z \prec y$. $x \prec z$ contradicts the fact that
$x \prec \cdot y$. $z \prec x$ implies $z \in A = \downarrow x$ which is, of course, not possible since
$z \in \overline{A}$ and $(A, \overline{A})$ is a D-cut. Hence $z \text{ co } x$. We have found a $z \in X$ such that
$x \neq z \neq y$ and $x \text{ co } z \prec y$, so that the proof is done in this case.
     We can assume now that $y \in Min(\overline{A})$. We know that $x \in Max(A)$. If
$x \notin Obmax(A)$ then, by Proposition 2.4.6(i), $\exists z' \in X \ \exists l \in L$ such that $z' \succ x$
and $l \cap [x, z'] = \{x\}$. If $z' \text{ co } y$ and $z' \neq y$ then the proof is done. $z' \prec y$
contradicts $x \prec y$. $y \preceq z'$ implies $y \notin l$. $y \notin l$ implies $\exists z \in l$ such that $z \text{ co } y$
and $z \neq y$; since $z \in l$ we have $z \succ x$ and $z \neq x$, and again the proof is done.
By analogy, if $y \notin Obmin(\overline{A})$ then the proof is also done. So the last possibility
is that both $x \in Obmax(A)$ and $y \in Obmin(\overline{A})$. But in this case $x, y \in c(A)$
and if we take any line $l' \supseteq \{x, y\}$ then we get $|c(A) \cap l'| = 2$ which contradicts
the hypothesis.

$(\Leftarrow)$ (By contradiction.) Suppose that $(X, \prec)$ is not jump-free. Then for
some $A \in D$ and $l \in L$, $|c(A) \cap l| = 2$. Let $c(A) \cap l = \{x, y\}$. Assume,

without loss of generality, that $x \prec y$. Clearly, $x \in Obmax(A)$, $y \in Obmin(\overline{A})$ and $x \prec\!\!\cdot\, y$. By the non-single degree property, $\exists z \neq y$ such that $x \prec z \; co \; y$ or $\exists z \neq x$ such that $x \; co \; z \prec y$. Assume the former (the latter case can be handled analogously). We have $l \cap [x, z] = \{x\}$ and by Proposition 2.4.6(i), $x \notin Obmax(A)$, a contradiction. $\square$ 2.4.10

It remains to characterise the case that $|c(A) \cap l| \neq 0$. To this end a connection to K-density can be made. Note first that if $c(A)$ contains a cut then K-density implies $|c(A) \cap l| \neq 0$. Conversely, it is easy to see that a poset is K-dense provided it is free of gaps. Hence we need an appropriate condition to ensure that $c(A)$ contains a cut. As it will turn out, this is achieved by the next definition.

Definition 2.4.11 [Strong cut-boundedness]
$(X, \prec)$ is strongly cut-bounded *iff* $\forall A \in D: A \subseteq \downarrow c(A) \wedge \overline{A} \subseteq \uparrow c(A)$. $\square$ 2.4.11

Theorem 2.4.12 [Characterisation of gap-freeness]
$(X, \prec)$ *has no gaps iff* $(X, \prec)$ *is K-dense and strongly cut-bounded.*

Proof:

($\Rightarrow$)   We prove: (a) $(X, \prec)$ is K-dense and (b) $(X, \prec)$ is strongly cut-bounded.
  (a) $(X, \prec)$ is K-dense.
  Let $c \in C$ and $l \in L$ be arbitrary; we have to prove that $c \cap l \neq \emptyset$. If $c = \downarrow c = \uparrow c$ then $|l| = 1$ and the result is trivially true. We can assume, without loss of generality, that $c \neq \uparrow c$ (the case that $c \neq \downarrow c$ is symmetrical).
  Define $A = \downarrow c$ and $\overline{A} = X \setminus A$. $(A, \overline{A})$ is clearly a D-cut and $Max(A) = c$, i.e., $Obmax(A) \subseteq c$. Since $(X, \prec)$ is free of gaps we get: $c(A) \cap l \neq \emptyset$. Let $x_0 \in c(A) \cap l$. If $x_0 \in Obmax(A) \subseteq c$ then $x_0 \in c$ and $c \cap l \neq \emptyset$. In this case the proof is done. Assume $x_0 \in Obmin(\overline{A})$. $x_0 \notin c$ implies $\exists y_0 \in c$ such that $y_0 \prec x_0$. But then by Proposition 2.4.6(ii) we have: $\exists w \in l \cap [y_0, x_0]$ such that $w \neq x_0$, i.e., $y_0 \preceq w \prec x_0$. $w \prec x_0$ implies $w \in A = \downarrow c$ and this implies $w = y_0$, i.e., $y_0 \in l$ and $c \cap l \neq \emptyset$. This proves Part (a).
  (b) $(X, \prec)$ is strongly cut-bounded.
  We have to prove that $\forall A \in D: A \subseteq \downarrow c(A) \wedge \overline{A} \subseteq \uparrow c(A)$. We prove only $A \subseteq \downarrow c(A)$ since the proof of the other part is completely similar. Let $A \in D$ and $x \in A$; we have to prove that $x \in \downarrow c(A)$. Let $l \in L$ be a line such that $x \in l$. Since $(X, \prec)$ is free of gaps we have:

$$\forall A \in D: c(A) \cap l \neq \emptyset.$$

Let $y \in c(A) \cap l$. Clearly $x \preceq y$, but $y \in c(A)$ which means that $x \in \downarrow c(A)$, and the proof is done. This proves Part (b).

($\Leftarrow$)   We have to prove that $\forall A \in D \; \forall l \in L : c(A) \cap l \neq \emptyset$. Let $A \in D$ and $l \in L$ be arbitrary. Clearly $c(A) \neq \emptyset$ since $(X, \prec)$ is strongly cut-bounded and $A \neq \emptyset$. If $\exists c \in C$ such that $c \subseteq c(A)$ then the result follows immediately since $(X, \prec)$ is K-dense. We shall prove that this is always the case.

Assume not, i.e., $\forall c \in C : c \not\subseteq c(A)$; we will arrive at a contradiction. Let $c' \subseteq c(A)$ be a maximal co-set in $c(A)$; c' exists and is non-empty because $c(A) \neq \emptyset$. Let $c \in C$ such that $c' \subseteq c$. Let $x \in c \setminus c'$; such an element x exists because $c \not\subseteq c(A)$. Assume that $x \in A$. (The case that $x \in \overline{A}$ is similar and we omit it.) $x \in A$ implies $x \in \downarrow c(A)$ since $(X, \prec)$ is strongly cut-bounded. $\exists z \in c(A)$ such that $x \prec z$ ($x \neq z$ since $x \in c$ and $c \not\subseteq c(A)$). $z \notin c'$ since x co x' for all $x' \in c'$. So, $\exists x' \in c' : z \; li \; x'$. $z \prec x'$ implies $x \prec x'$ which is a contradiction. Hence $x' \prec z$.

Then $x' \in Obmax(A)$ and $z \in Obmin(\overline{A})$. $z \in Obmin(\overline{A})$ and $x' \prec z$ imply, by Proposition 2.4.6(ii), that for all $l' \in L$ such that $z \in l'$ we have $l' \cap [x', z] \neq \{z\}$. But this means that for all $l' \in L$ such that $z \in l'$ we have $x' \in l'$ since between x' and z there is no element different from x' and z. In particular, if $l \in L$ is a line such that $x, z \in l$ then by the above argument $x' \in l$, i.e., $x \; li \; x'$ which is a contradiction since $x \neq x'$ and x co x'. This contradiction proves the original claim that $c(A)$ contains a cut, and with it the theorem.   □ 2.4.12

From Theorems 2.4.10, 2.4.12 and the fact that a poset is D-continuous iff it is free of jumps and free of gaps, a first characterisation of D-continuity already follows. However, strong cut-boundedness remains a rather complicated property, relying as it does on the definition of $c(A)$. We define a property next which is related to strong cut-boundedness and which can be used to characterise strong cut-boundedness in simpler terms. However, the details of this characterisation will be relegated to the exercises because they are not needed later in this book (see Exercises 25 and 26).

Definition 2.4.13 [Cut-boundedness]

(i) $(X, \prec)$ is cut-bounded from below *iff* $\forall A \in D : \overline{A} \subseteq \uparrow M(A)$.

(ii) $(X, \prec)$ is cut-bounded from above *iff* $\forall A \in D : A \subseteq \downarrow M(A)$.

(iii) $(X, \prec)$ is cut-bounded *iff* $\forall A \in D : A \subseteq \downarrow M(A) \wedge \overline{A} \subseteq \uparrow M(A)$.
    □ 2.4.13

In Section 2.5, it will be shown that in interesting special cases, strong cut-boundedness and (plain) cut-boundedness denote the same properties.

We will now state the main result of this section, a characterisation of D-continuity by means of three independent properties. The fact that these properties are independent is shown in Figure 2.21 which exhibits a few typical examples relating to the concepts introduced in this section.

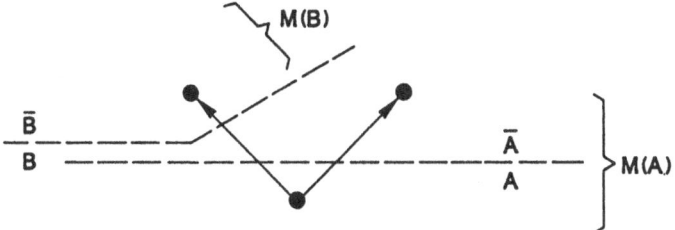

(i) D-continuous, K-dense, non-single degree, strongly cut-bounded

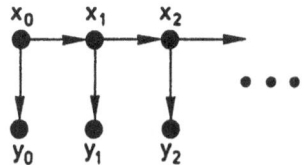

(ii) ¬ D-continuous, ¬ K-dense, non-single degree, strongly cut-bounded

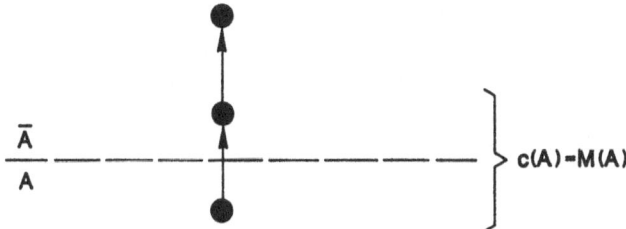

(iii) ¬ D-continuous, K-dense, ¬ non-single degree, strongly cut-bounded

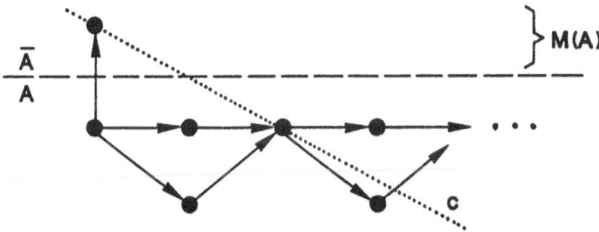

(iv) ¬ D-continuous, K-dense, non-single degree, ¬ cut-bounded,¬ strongly cut-bounded

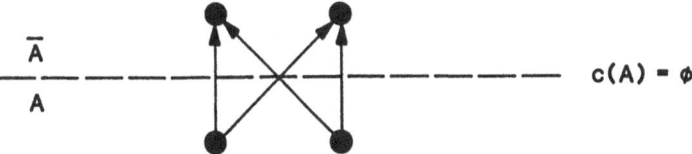

(v) ¬ D-continuous, K-dense, non-single degree, cut-bounded,¬ strongly cut-bounded

Fig. 2.21. Some examples relating to D-continuity

Theorem 2.4.14 [Characterisation of D-continuity]
A poset $(X, \prec)$ is D-continuous if and only if it is K-dense, of non-single degree and strongly cut-bounded.

Proof: From Theorem 2.4.12, Theorem 2.4.10 and the fact that $(X, \prec)$ is D-continuous iff it has neither jumps nor gaps.   □ 2.4.14

2.5   Occurrence Posets

In this section we shall study a particular type of posets called occurrence posets. Most of the results of the previous sections can be particularised nicely to this type of poset.

The study of occurrence posets has been motivated by the study of processes modelled by occurrence nets. In the definition of occurrence posets, we will use the symbols B and E that will be interpreted later when occurrence nets are introduced. However, being consequent with the mathematical nature of this work – going from general structures to particular ones – we shall delay the definition and the interpretation of occurrence nets to the next chapter.

Definition 2.5.1 [Occurrence posets]
A poset $(X, \prec)$ will be called an occurrence poset *iff*

1. $(X, \prec)$ is combinatorial.

2.   There exists a partitioning $X = B \uplus E$ of X such that

(i) $\prec \; \subseteq (B \times E) \cup (E \times B)$,

(ii) $\forall b \in B: |{}^\bullet b| \leq 1 \wedge |b^\bullet| \leq 1$.   □ 2.5.1

Figure 2.22(a) shows an example of an infinite occurrence poset. Figure 2.22(b) shows an occurrence poset such that $E = \{e_1, e_0, e_2\}$ and $B = X \backslash E$. Figure 2.22(c) shows a combinatorial poset which is not an occurrence poset. It violates Definition 2.5.1 because the two elements e_1 and e_2 should be included in the set E to satisfy Requirement 2.5.1.2(ii), but then there is no way to define B such that Requirement 2.5.1.2(i) is satisfied. An 'even' $\prec$ -chain like that from e_1 to e_2 in Figure 2.22(b) can be tolerated, while a similar 'odd' chain like that in Figure 2.22(c) should be disallowed. This can be formalised as follows.

Definition 2.5.2 [B-likeness]
Let $(X, \prec)$ be a poset and $x \in X$.
x will be called B-like *iff* $|{}^\bullet x| \leq 1 \wedge |x^\bullet| \leq 1$.   □ 2.5.2

Between non-B-like elements, only 'even' $\prec$ -chains may be allowed. Thus, we arrive at the following characterisation of occurrence posets:

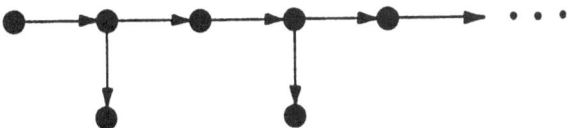

(a) An infinite occurrence poset

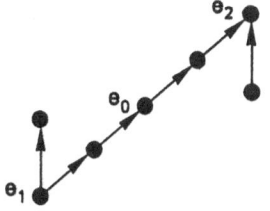

(b) A finite occurrence poset

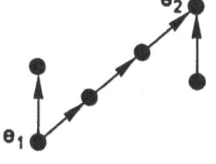

(c) Not an occurrence poset

Fig. 2.22. Illustration of occurrence posets

Theorem 2.5.3 [Characterisation of occurrence posets]

$(X, \prec)$ *is an occurrence poset* $\Longleftrightarrow$

(i) $(X, \prec)$ *is combinatorial, and*

(ii) $\forall x, y \in X : (x \prec^q y$ *and* q *is odd and* x *is not B-like*) $\Rightarrow y$ *is B-like.*

Proof:

($\Rightarrow$)   An occurrence poset is combinatorial by definition. Furthermore, for all partitionings $X = B \uplus E$, all non-B-like elements must be in E by Definition 2.5.1.2(ii), and a $\prec$ -chain of odd length between any two such elements contradicts Definition 2.5.1.2(i).

($\Leftarrow$)   We have to define appropriate subsets B and E of X. First, we may assume $(X, \prec)$ to be li-connected, that is, by definition, $X \times X = li^*$; if this is not true then the li-components of $(X, \prec)$ may be considered separately.
Then there are two cases: either all elements of X are B-like, or at least one of them is not.
In the first case, it follows easily that $(X, \prec)$ is a total order (i.e., $co = id|_X$) of only B-like elements. We may choose any one of these elements as e_0, and define

$E = \{x \in X \mid \exists n \in \mathbf{N} : n$ is even and $(e_0, x) \in (\prec \cup \prec^{-1})^n\}$
$B = X \backslash E.$

It is easy to prove Items 2(i) and 2(ii) of Definition 2.5.1 in this case.
In case there are some non-B-like elements, we may take any one of them as e_0 and define E and B in just the same way as before. Then it has to be shown that the resulting partitioning of X satifies the properties 2.5.1.2(i,ii) of

an occurrence poset. First, we prove the auxiliary fact that the existence of an odd $(\prec \cup \prec^{-1})$-chain from e_0 to $x \in X$ implies that x is B-like; i.e.:

$$(e_0, x) \in (\prec \cup \prec^{-1})^q, \ q \text{ odd} \ \Rightarrow \ x \text{ is } B-\text{like}.$$

To this end, assume that there are elements $x_0, x_1, \ldots, x_{q-1}, x_q$ such that q is odd, $x_0 = e_0$, $x_q = x$ and $(x_i, x_{i+1}) \in (\prec \cup \prec^{-1})$ for $0 \le i < q$. We prove the claim by induction on q. If $q = 1$ then either $e_0 \prec \cdot x$ or $x \prec \cdot e_0$; in both cases, the claim that x is B-like follows from the non-B-likeness of e_0 and Premise (ii) of the theorem. Suppose, on the other hand, that $q > 1$. We may distinguish two cases: $x_0 \prec \cdot x_1$ and $x_1 \prec \cdot x_0$. Suppose first that $x_0 \prec \cdot x_1$. If $\forall i, 0 \le i < q: (x_i \prec x_{i+1})$ then, again, the claim follows from Premise (ii) of the theorem. Otherwise, there is an index k such that $0 \le k < q$ and $x_{k+1} \prec \cdot x_k$ and $\forall i, 0 \le i < k: x_i \prec x_{i+1}$. If $x_{k-1} = x_{k+1}$ then the sequence

$$x_0, x_1, \ldots, x_{k-1}(= x_{k+1}), \ \ldots, x_{q-1}, x_q$$

is again a $(\prec \cup \prec^{-1})$-path of odd length (shorter than q) from e_0 to x, and the claim that x is B-like follows by the non-B-likeness of e_0 and the induction hypothesis. If, on the other hand, $x_{k-1} \ne x_{k+1}$, then x_k is a non-B-like element (since $|^{\bullet}x_k| \ge |\{x_{k-1}, x_{k+1}\}| = 2$), and by Premise (ii) of the theorem, k must be an even number ≥ 2. Hence $x_k, x_{k+1}, \ldots, x_{q-1}, x_q$ is again a $(\prec \cup \prec^{-1})$-path of odd length (shorter than q) from a non-B-like element (namely x_k) to x, and the claim follows again from the induction hypothesis. If $x_1 \prec \cdot x_0$ then the claim can be proved analogously.

Now we turn to the proof of Property 2(i) of Definition 2.5.1; we do this proof in two parts: first we show that $x \prec y \wedge x \in B$ implies $y \in E$, and then we show that $x \prec y \wedge x \in E$ implies $y \in B$.

Assume that $x \prec y$ and $x \in B$. By li-connectedness we have $(e_0, x) \in li^*$, and by combinatorialness, this implies $(e_0, x) \in (\prec \cup \prec^{-1})^*$.

But $x \in B$, hence $x \notin E$, and the $(\prec \cup \prec^{-1})$-chain that leads from e_0 to x can only be of odd length; hence since $x \prec y$, some even $(\prec \cup \prec^{-1})$-chain leads from e_0 to y, and $y \in E$ by the definition of E.

Assume, on the other hand, that $x \prec y$ and $x \in E$. By definition, an even $(\prec \cup \prec^{-1})$-chain leads from e_0 to x, and we have to prove that no even $(\prec \cup \prec^{-1})$-chain at all leads from e_0 to y. But suppose there does; then some odd $(\prec \cup \prec^{-1})$-chain also leads from e_0 to x, since $x \prec y$. Hence x as well as y must be B-like.

Similarly, all elements on the two chains, up to and including e_0, must be B-like; however, this contradicts the assumption that e_0 is not B-like. Hence y is not in E, which means that it must be in B, and this finishes the proof of Property 2.5.1.2(i).

Finally, it remains to prove Property 2.5.1.2(ii). Suppose that $b \in B$; then, as before, an odd $(\prec \cup \prec^{-1})$-chain leads from e_0 to b, and b must be B-like.

$\square$ 2.5.3

A nice property of occurrence posets is the fact that they are always N-dense. This is the result of our next theorem.

Theorem 2.5.4 [Occurrence posets are N-dense]
Let $(X, \prec)$ be an occurrence poset. Then $(X, \prec)$ is N-dense.

Proof: Let x, y, x', y' be elements of X such that:

$$x \prec y \wedge x \prec x' \wedge y' \prec y \wedge (x \text{ co } y' \text{ co } x' \text{ co } y).$$

We have to prove that

$$\exists z \in X : x \prec z \prec y \wedge (x' \text{ co } z \text{ co } y').$$

Since $(X, \prec)$ is an occurrence poset it is combinatorial by Definition 2.5.1.1. Thus, $\exists x_0, x_1, \ldots, x_n$ such that:

$$x = x_0 \prec x_1 \prec \ldots x_{n-1} \prec x_n = y.$$

Let $1 \leq j \leq n$ be the smallest natural number for which x_j co x' (j clearly exists since $y = x_n$ co x'). If x_j co y' then x_j is the z we are looking for and the proof is done.
    Claim: x_j co y'.
First note that $x_j \prec y'$ is not possible since in this case $x_0 \prec x_j \prec y'$ which is a contradiction with the fact that x_0 co y'. Assume that $y' \prec x_j$ (see Figure 2.23).

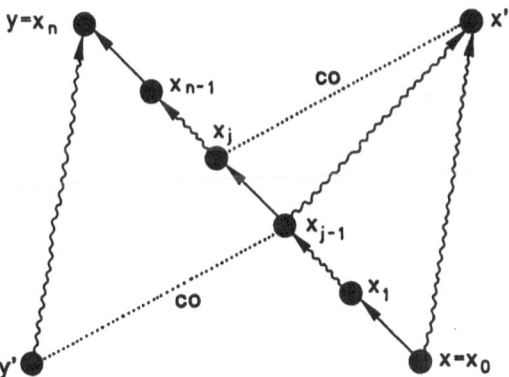

Fig. 2.23. Illustration of the proof that occurrence posets are N-dense

Now, by the definition of j, x_{j-1} *li* x' and since $x' \preceq x_{j-1}$ is not possible (because in that case $x' \preceq x_{j-1} \prec y$, a contradiction since x' co y), then we must have $x_{j-1} \prec x'$. $y' \prec x_{j-1}$ implies $y' \prec x_{j-1} \prec x'$, a contradiction since y' co x'. $x_{j-1} \prec y'$ implies $x \preceq x_{j-1} \prec y'$, a contradiction since x co y'. It follows that y' co x_{j-1}. Now, x_{j-1} is clearly not B-like. Then, by Theorem 2.5.3(ii) we must have that x_j is B-like, a contradiction with the assumption $y' \prec x_j$ and with the fact that y' co x_{j-1}. It follows that y' co x_j. So, the claim and the theorem are proved.   □ 2.5.4

The results about K-density can be specialised as follows:

Theorem 2.5.5 [Characterisations of K-density for occurrence posets]
Let $(X, \prec)$ be an occurrence poset with a partitioning $X = B \uplus E$.

(i) *$(X, \prec)$ is K-dense iff none of the two occurrence posets shown in Figure 2.24 can be embedded into it.*

(ii) *$(X, \prec)$ is K-dense iff it is weakly discrete and satisfies the line crossing property.*

Proof: Since an occurrence poset is combinatorial by definition and N-dense by Theorem 2.5.4, Theorem 2.3.6 can be applied. In Figure 2.24, the posets shown in Figure 2.10 have been modified to make occurrence posets out of them. However, Theorem 2.3.6 is not affected because, as will be shown next, if the poset shown in Figure 2.10(i) is embeddable into an occurrence poset then so is the poset shown in Figure 2.24(i), and symmetrically for Figures 2.10(ii) and 2.24(ii).

Suppose that the poset $(X', \prec') = (\{x_1, x_2, \ldots\} \cup \{y_1, y_2, \ldots\}, \prec')$ of Figure 2.10(i) (with the labelling given there) can be embedded into $(X, \prec)$, with an injection γ as in Definition 2.3.5. First we claim that without loss of generality, all $\gamma(x_i)$ can be assumed to lie in the set E. Assume, to the contrary, that $\gamma(x_k) \in B$; then $\gamma(x_k)^\bullet \neq \emptyset$ because $(X', \prec')$ is infinite, and by $\gamma(x_k) \in B$, $|\gamma(x_k)^\bullet| = 1$. Then for $e \in \gamma(x_k)^\bullet$, the function γ' with $\gamma'(x_k) = e$, $\gamma'(z) = \gamma(z)$ for $z \neq x_k$ satisfies, as is easy to verify, the same embeddability properties as does γ.

So, assume $\gamma(\{x_1, x_2, \ldots\}) \subseteq E$ and consider an index $i \geq 1$. Because of $\gamma(x_i) \prec \gamma(x_{i+1})$ and $\gamma(x_i), \gamma(x_{i+1}) \in E$, there is an F-chain

$$w_0^{(i)} \; F \; w_1^{(i)} \; F \; \ldots \; F \; w_m^{(i)}$$

with $m \geq 2$ and $w_0^{(i)} = \gamma(x_i)$, $w_m^{(i)} = \gamma(x_{i+1})$. With $\gamma''(z_i) = w_{m-1}^{(i)}$ and $\gamma'' = \gamma$ for all x_i and y_i, γ'' is a function that satisfies the embeddability properties for the poset shown in Figure 2.24(i).   □ 2.5.5

We now turn to the characterisation of Dedekind continuity derived in Section 2.4, Theorem 2.4.14. The next theorems show that two of the characterising properties can be simplified for occurrence posets.

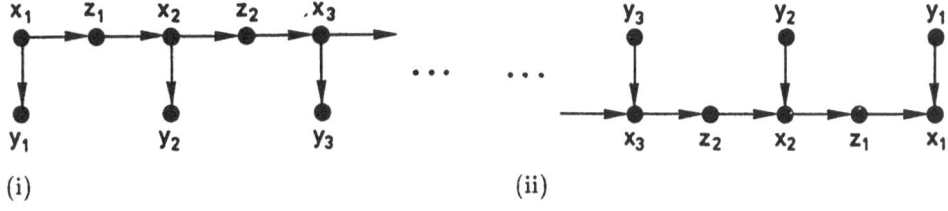

Fig. 2.24. Two non-K-dense occurrence posets

Theorem 2.5.6 [The non-single degree property for occurrence posets]
Let $(X, \prec)$ be an occurrence poset with a partitioning $X = B \uplus E$.
$(X, \prec)$ satisfies the non-single degree property iff $\forall e \in E: |{}^\bullet e| \neq 1 \neq |e^\bullet|$.

Proof:

($\Rightarrow$)   Let $e \in E$. We prove that $|{}^\bullet e| \neq 1$.

If $|{}^\bullet e| = 0$ then we are done. Assume $|{}^\bullet e| \neq 0$, i.e., ${}^\bullet e \neq \emptyset$. We have to prove that $|{}^\bullet e| > 1$. Let $x \in {}^\bullet e$. Since $(X, \prec)$ is of non-single degree,

$$\exists z \in X: x \neq z \neq e: (x \prec z \; co \; e) \vee (x \; co \; z \prec e).$$

Assume $x \prec z \; co \; e$ (see Figure 2.25).

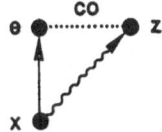

Fig. 2.25. Illustration of the case that $x \prec z \; co \; e$

We have $(x, e) \in \prec \; \subseteq (B \times E) \cup (E \times B)$. Then $(x, e) \in B \times E$ and $x \in B$, a contradiction since $|x^\bullet| > 1$. So, we must have $x \; co \; z \prec e$ (see Figure 2.26).

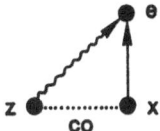

Fig. 2.26. Illustration of the case that $x \; co \; z \prec e$

But in this case we have $|^{\bullet}e| > 1$ which was to be proved.
The proof that $|e^{\bullet}| \neq 1$ is similar and we omit it.

($\Leftarrow$)   Let $x, y \in X$ be such that $x \prec y$. We have to show that

$$\exists z \in X: x \neq z \neq y: (x \prec z \text{ co } y) \vee (x \text{ co } z \prec y).$$

$(x, y) \in (B \times E) \cup (E \times B)$ implies that $(x, y) \in B \times E$ or $(x, y) \in E \times B$. If $(x, y) \in B \times E$ then, since $|^{\bullet}y| \neq 1$ by hypothesis, we have

$$\exists z: x \neq z \neq y: x \text{ co } z \prec y,$$

and we are done. If $(x, y) \in E \times B$ then, since $|x^{\bullet}| \neq 1$ by hypothesis, we have

$$\exists z: x \neq z \neq y: x \prec z \text{ co } y,$$

and we are again done. $\square$ 2.5.6

Theorem 2.5.7
Let $(X, \prec)$ be an occurrence poset.
Then $(X, \prec)$ is strongly cut-bounded iff it is cut-bounded.

Proof: Since $c(A) \subseteq M(A)$ by definition, it follows trivially that if $(X, \prec)$ is strongly cut-bounded then $(X, \prec)$ is cut-bounded.
   We only have to prove the other direction, i.e., we have to prove that $\forall A \in D: A \subseteq \downarrow c(A) \wedge \overline{A} \subseteq \uparrow c(A)$. We shall prove only that $A \subseteq \downarrow c(A)$; the proof that $\overline{A} \subseteq \uparrow c(A)$ is similar.
   Let $(A, \overline{A}) \in D$ and $x \in A$. By hypothesis, $A \subseteq \downarrow M(A)$. This means that $\exists y \in M(A) = Max(A) \cup Min(\overline{A})$ such that $x \prec y$.

Case 1: $y \in Max(A)$.
If $|y^{\bullet}| = 0$ then clearly $y \in Obmax(A) \subseteq c(A)$ and the proof is done. So, we can assume that $|y^{\bullet}| \neq 0$, i.e., $y^{\bullet} \neq \emptyset$.
   Let $z \in y^{\bullet}$.
Since $y \in Max(A)$, it follows that $z \in \overline{A}$. Now, $(y, z) \in \prec \subseteq (B \times E) \cup (E \times B)$.
If $(y, z) \in B \times E$ then $y \in B$ and by Corollary 2.4.7(i), $y \in Obmax(A) \subseteq c(A)$.
If $(y, z) \in E \times B$ then $z \in B$ and for the same reasons as before, $z \in Min(\overline{A})$ and, by Corollary 2.4.7(ii), $z \in Obmin(\overline{A}) \subseteq c(A)$. In both cases the proof is finished.

Case 2: $y \in Min(\overline{A})$.
Since $x \prec y$ and $(X, \prec)$ is combinatorial we have $\exists x_0, x_1, \ldots, x_n$ such that:

$$x = x_0 \prec x_1 \prec \ldots \prec x_n = y.$$

Since $y \in Min(\overline{A})$ then $x_{n-1} \in A$. Again $(x_{n-1}, y) \in \prec \subseteq (B \times E) \cup (E \times B)$.
For the same reasons as before, it follows that either $y \in Obmin(\overline{A}) \subseteq c(A)$ or $x_{n-1} \in Obmax(A) \subseteq c(A)$.

   In both cases the proof is again done. $\square$ 2.5.7

Theorem 2.5.8 [Characterisation of D-continuity for occurrence posets]
*Let $(X, \prec)$ be an occurrence poset with a partitioning $X = B \uplus E$.
Then $(X, \prec)$ is D-continuous $\iff$
$(X, \prec)$ is K-dense, cut-bounded and $\forall e \in E: |{}^\bullet e| \neq 1 \neq |e^\bullet|$.*

Proof: From Theorem 2.4.14, together with Theorems 2.5.6 and 2.5.7.

$\square$ 2.5.8

Exercises.

1. Calculate all cuts and all lines of the posets shown in Figures 2.1, 2.4, 2.6(i) and 2.6(ii).

2. We can define two order relations on the set of all cuts of a poset:

   Definition

   Let $(X, \prec)$ be a poset and $c_1, c_2 \in C(X, \prec)$.
   $c_1 \sqsubseteq_1 c_2$ *iff* $\forall x \in c_1 \; \exists y \in c_2 : x \preceq y$;
   $c_1 \sqsubseteq_2 c_2$ *iff* $\forall y \in c_2 \; \exists x \in c_1 : x \preceq y$. $\square$

   (i)   Check that $\sqsubseteq_1$ and $\sqsubseteq_2$ are partial order relations on $C(X, \prec)$.

   (ii)   Prove that $c_1 \sqsubseteq_1 c_2 \iff c_1 \sqsubseteq_2 c_2$. Having proved this, we may identify $\sqsubseteq_1$ and $\sqsubseteq_2$ to equal, by definition, $\sqsubseteq$.

   (iii)   Calculate the relation $\sqsubseteq$ between the cuts of the posets shown in Figure 2.1.

   (iv)   Find a poset with two cuts c_1, c_2 such that neither $c_1 \sqsubseteq c_2$ nor $c_2 \sqsubseteq c_1$.

3. What are the immediate neighbours of the elements of the posets shown in Figures 2.1, 2.4 and 2.27?

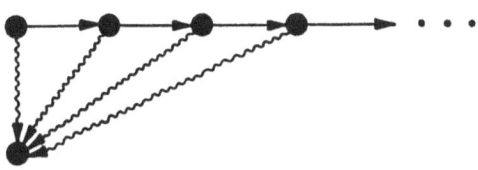

Fig. 2.27. A linearly ordered poset

4. Is the poset shown in Figure 2.27

    (i)   dense?

    (ii)   combinatorial?

5. Prove or disprove:
$(X, \prec)$ combinatorial $\Rightarrow (X, \prec)$ weakly discrete.

6. (i)   Find a poset which is combinatorial and of finite degree but not weakly discrete.

    (ii)   Show that in Theorem 2.2.10(iii), the assumption of weak discreteness cannot be relaxed to combinatorialness.

7. Show that both assumptions in Theorem 2.2.12 are necessary (and independent of each other).

8. For the poset shown in Figure 2.28, define

    (i)   an observer according to Theorem 2.2.15(ii) and

    (ii)   an injective observer according to Theorem 2.2.16($\Leftarrow$),

taking the enumeration as shown in the figure.

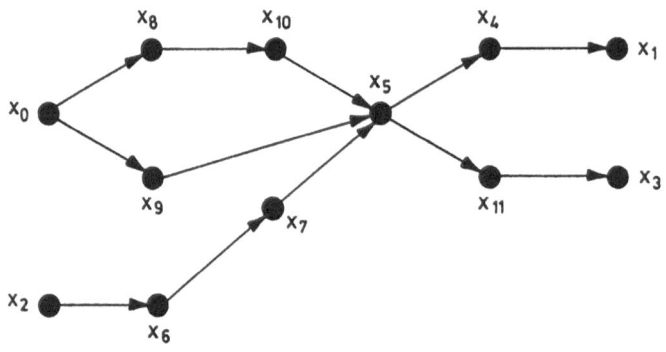

Fig. 2.28. Calculation of observers

9. Define a poset to be observable with respect to $\mathbf{N}$ *iff*
$\exists f\colon X \to \mathbf{N}\colon \forall x, y \in X\colon x \prec y \Rightarrow f(x) < f(y)$.
Prove that $(X, \prec)$ is observable w.r.t. $\mathbf{N}$ iff $\forall x \in X\colon |\downarrow x| \in \mathbf{N}$.

10.   (i)   Show that the construction in the proof of Theorem 2.2.15(ii) does not yield *all* possible observers.

∗ (ii)   Show that the following more general construction
(a) yields an observer; (b) yields all possible observers.
Define $N_1^i, N_2^i \in (\mathbf{Z} \cup \{-\infty, +\infty\})$, $f(x_i) \in \mathbf{Z}$ as follows:

$$N_1^i = \begin{cases} \max_{k \in \alpha_i} \{f(x_k) + \nu(x_k, x_i) - 1\} & \text{if } \alpha_i \neq \emptyset \\ -\infty & \text{if } \alpha_i = \emptyset \end{cases}$$

$$N_2^i = \begin{cases} \min_{j \in \beta_i} \{f(x_j) - \nu(x_i, x_j) + 1\} & \text{if } \beta_i \neq \emptyset \\ +\infty & \text{if } \beta_i = \emptyset \end{cases}$$

Choose for $f(x_i)$ any integer in the set $\mathbf{Z} \cap [N_1^i, N_2^i]$. (It must be proved that always $N_1^i \leq N_2^i$.)

11. ∗ Find an uncountable poset which is boundedly discrete but not observable.

12. Show that the construction in Theorem 2.2.16($\Leftarrow$) may yield all possible injective observers of a given poset.

13. Give a construction which allows to find all possible observers with respect to a given cut c according to Theorem 2.2.17($\Leftarrow$).

14. Check whether the posets shown in Figures 2.1, 2.4, 2.6(i,ii) and 2.27 are:

   (i)   N-dense;

   (ii)   K-dense.

15. Invent a poset which is dense but not N-dense.

16. Prove or disprove:
    Let $(X, \prec)$ be a combinatorial poset.
    Then $(X, \prec)$ is N-dense iff $\forall x, y \in X : (x^\bullet \cup y^\bullet \neq \emptyset \Rightarrow x^\bullet = y^\bullet)$.

17. Prove or disprove:

   (i)   K-density $\Rightarrow$ line crossing property;

   (ii)   N-density $\Rightarrow$ line crossing property.

18. Is the poset shown in Figure 2.10(i) embeddable, respectively, into the posets shown in Figures 2.4, 2.27 and your result of Exercise 6(i)?

19. (i)   Show that Theorem 2.3.6((i) $\Rightarrow$ (ii)) fails to hold if combinatorialness is omitted from the list of assumptions; i.e., find a poset which is K-dense but not combinatorial (and hence, by Theorem 2.2.6, not weakly discrete).

   (ii)   Show that Theorem 2.3.6((iii) $\Rightarrow$ (i)) fails to hold if N-density is not assumed; i.e., find a poset into which neither of the two posets shown in Figure 2.10(i,ii) is embeddable and which is not N-dense (and hence not K-dense).

20. Prove that all four assumptions in Theorem 2.3.12 are necessary.

21. Calculate all D-cuts for the posets shown in Figures 2.1, 2.27, 2.6(ii) and 2.21(i,iii,v).

22. Prove or disprove:

    (i)   If $(A, \overline{A})$ is a D-cut of $(X, \prec)$ then $A = \downarrow A$ and $\overline{A} = \uparrow \overline{A}$.

    (ii)   Let $A \subseteq X$ such that $\emptyset \neq A \neq X$. Then if $A = \downarrow A$ then the pair $(A, \overline{A})$ where $\overline{A} = X \setminus A$ is a D-cut.

23. Calculate $M(A)$ and $c(A)$

    (i)   for all D-cuts of the posets shown in Figures 2.1 and 2.21(i,iii,iv);

    (ii)   for several other D-cuts of your choice.

24. Are the posets shown in Figures 2.21(i-v) jump-free and gap-free, respectively?

25. * Prove the following technical fact about D-cuts:
Let $(X, \prec)$ be a poset, $A \in D$, $X_0 \subseteq Max(A)$, $X_0 \neq A$ and define $B = A \setminus X_0$, $\overline{B} = X \setminus B$.
Then:

    (i) $B \in D$, i.e., $B \neq \emptyset \neq \overline{B}$ and $\downarrow B = B$.

    (ii) $X_0 \subseteq Min(\overline{B})$ and $X_0 \neq \overline{B}$.

    (iii)   With

$$Y_1 = \{y \in A \mid \exists x \in X_0 : y \prec x\}$$
$$Y_2 = \{y \in \overline{A} \mid \exists x \in X_0 : x \prec y\}$$

we have:

$$M(A) \setminus Y_2 = M(B) \setminus Y_1.$$

26. * Using this fact, prove that the 'fork property' defined below describes the difference between cut-boundedness and strong cut-boundedness. That is, prove that a poset $(X, \prec)$ is strongly cut-bounded iff it is cut-bounded and satisfies both the up-fork property and the low-fork property.

Definition.

(i) $(X, \prec)$ has the up-fork property *iff*
$\forall x, y \in X,\ x \prec y\colon \forall l \in L\colon l \cap [x, y] = \{x\} \ \Rightarrow\ (\exists y' \succ x\colon \downarrow y' = \downarrow x \cup \{y'\})$;

(ii) $(X, \prec)$ has the low-fork property *iff*
$\forall x, y \in X,\ x \prec y\colon \forall l \in L\colon l \cap [x, y] = \{y\} \ \Rightarrow\ (\exists x' \prec y\colon \uparrow x' = \uparrow y \cup \{x'\})$.
$\square$

Chapter 3. Petri Nets

3.1 Nets and Markings

We have motivated the study of partially ordered sets by the claim that they
may be used to define concurrent behaviour faithfully. But we have not yet
specified the systems the behaviours of which are to be described.

As illustrated in Chapter 1 of this book, Petri net theory stipulates that
a concurrent system be described by two kinds of objects, state-like objects S
(sometimes called places) and action-like objects T (sometimes called transitions). The idea is that state objects may hold to make up a certain state, and
that transitions may occur and thus change the state. Thus, state objects and
transitions are interconnected in an alternating fashion, which in net theory is
captured by an interconnection relation F (for flow) defined as a subset of the
set $(S \times T) \cup (T \times S)$.

Definition 3.1.1 [Nets]
(S, T, F) is a net *iff* $S \cap T = \emptyset$ and $F \subseteq (S \times T) \cup (T \times S)$. □ 3.1.1

We shall only consider non-empty nets, excluding the trivial case $S = T = \emptyset$
from now on.

The S-elements of a net are represented pictorially by circles, the T-elements
are represented by boxes, and the F relation is represented by arrows either
from circles to boxes or from boxes to circles. For examples illustrating nets,
the reader is referred to Chapter 1.

Some useful notation is given by the following:

Notation 3.1.2 [The 'dot' notation]
For $x \in S \cup T$,

$$^{\bullet}x = \{y \in S \cup T \mid (y, x) \in F\} \quad (\text{the pre} - \text{set of } x),$$
$$x^{\bullet} = \{y \in S \cup T \mid (x, y) \in F\} \quad (\text{the post} - \text{set of } x).$$

For $Y \subseteq S \cup T$,

$$^{\bullet}Y = \bigcup_{x \in Y} {}^{\bullet}x,$$
$$Y^{\bullet} = \bigcup_{x \in Y} x^{\bullet}.$$

□ 3.1.2

Since the interpretation will be that transitions represent changes-of-state, it introduces some difficulties to allow transitions t with $^\bullet t = \emptyset$ or transitions t with $t^\bullet = \emptyset$ or both. We will, first of all, exclude this by means of requiring

$$T \subseteq dom(F) \cap cod(F)$$

for every net $N = (S, T, F)$. By this convention, we lose nothing because all nets we shall be interested in have this property; but we shall avoid some technical problems.

From a formal point of view it is advantageous to interpret the relation F as a function

$$F : ((S \times T) \cup (T \times S)) \rightarrow \{0, 1\}$$

with the convention $(x, y) \in F \Leftrightarrow F(x, y) = 1$ and $(x, y) \notin F \iff F(x, y) = 0$. We will sometimes make use of this view in order to shorten formulae.

The places of a net describing a system may be marked by 'tokens' to indicate the state the system is in. Token distributions may change by the occurrence of transitions, representing a change of state governed by the transition rule. An example of this has been discussed in Chapter 1 (Figures 1.2 and 1.3). In general, state objects are allowed to hold several times, for example .representing the presence of several similar resources or counting the number of some items (like messages in the example of the introduction).

A place/transition net is a net together with an initial marking, describing the structure and the initial state, respectively, of a concurrent system. The underlying structure of a place/transition net is simply a net $N = (S, T, F)$ as introduced in Definition 3.1.1.

Definition 3.1.3 [System nets or place/transition nets]

(i) M is a marking of a net (S, T, F) *iff* $M : S \rightarrow \mathbf{N}$.

(ii) $\Sigma = (S, T, F, M_0)$ is a system net or place/transition net *iff* (S, T, F) is a net and M_0 is a marking (called the initial marking of Σ).
$$\square \; 3.1.3$$

We shall use the terms system net and place/transition net interchangeably. M_0 represents the initial state of the system described by Σ. Henceforth we will only consider countable system nets, i.e., such that $S \cup T$ is a countable set.

3.2   Transition Rule and Occurrence Sequences

The transition rule specifies under which conditions a marking M enables a transition t, and how the occurrence of t changes M into a new marking M'.

Definition 3.2.1 [Transition rule]

Let (S, T, F) be a net, M a marking and $t \in T$.

(i) M enables t *iff* $\forall s \in S \colon F(s, t) \leq M(s)$.

(ii) M' is produced from M by the occurrence of t (in symbols: $M[t\rangle M'$)
      iff M enables t and $\forall s \in S \colon M'(s) = M(s) - F(s, t) + F(t, s)$.   □ 3.2.1

This definition ensures that M' is again a marking, since if M enables t
then $M(s) \geq F(s, t)$, and the latter implies that $M(s) - F(s, t) + F(t, s)$ is
nonnegative.

Definition 3.2.2 [Occurrence sequences]

Let $\Sigma = (S, T, F, M_0)$ be a system net, let $M_1, M_2, \ldots$ be markings and let
$t_1, t_2, \ldots$ be transitions in T.

(i) $\sigma = M_0 t_1 M_1 \ldots t_n M_n$ is a (finite) occurrence sequence of Σ *iff*

$$\forall i, 1 \leq i \leq n \colon M_{i-1}[t_i\rangle M_i;$$

$\sigma = M_0 t_1 M_1 t_2 \ldots$ is an (infinite) occurrence sequence of Σ *iff*

$$\forall i, 1 \leq i \colon M_{i-1}[t_i\rangle M_i.$$

(ii) By $first(\sigma)$ we denote the first marking of $\sigma = M_0 t_1 M_1 \ldots$, i.e., M_0. In
      case $\sigma = M_0 t_1 \ldots t_n M_n$ is finite then $last(\sigma) = M_n$ is also defined.
      If σ is finite and σ' is an occurrence sequence of $(S, T, F, last(\sigma))$ then the
      catenation $\sigma'' = \sigma \sigma'$ is defined (by identifying $last(\sigma)$ and $first(\sigma')$) and
      is again an occurrence sequence of Σ; σ is called a prefix of σ''.

(iii) $[M_0\rangle = \{M \mid \exists \text{ occurrence sequence } \sigma \colon first(\sigma) = M_0 \wedge last(\sigma) = M\}$
      (the set of markings reachable from M_0 by successive occurrences of single
      transitions).   □ 3.2.2

Remark 3.2.3 [Transition sequences]

In the literature, instead of the occurrence sequences defined in Definition 3.2.2(i),
one may find the definition of 'firing sequences', that is, the restrictions of occurrence sequences to sequences of transitions:

$$\sigma = M_0 t_1 M_1 t_2 M_2 t_3 \ldots \quad \longrightarrow \quad \sigma_T = t_1 t_2 t_3 \ldots$$

We will call σ_T the *transition sequence* associated to σ. It may easily be seen
that from a restricted occurrence sequence σ_T, the initial marking M_0 and the
flow relation F, one may uniquely reconstruct the corresponding occurrence
sequence σ. One may also consider the restriction of σ to marking sequences:

$$\sigma \longrightarrow \sigma_M = M_0 M_1 M_2 \ldots,$$

but it is not possible to reconstruct σ uniquely from σ_M in general.   □ 3.2.3

58   3. Petri Nets

In an occurrence sequence $\sigma = M_0 t_1 M_1 t_2 \ldots$ we may, of course, have $t_i = t_j$ and $i \neq j$ as well as $M_i = M_j$ and $i \neq j$. That is, a transition t or a marking M may occur more than once in σ.

Pictorially, markings M are represented by placing $M(s)$ 'tokens' (black dots) on the place s. The occurrence of a transition can be visualised by tokens 'flowing' through the net according to the transition rule.

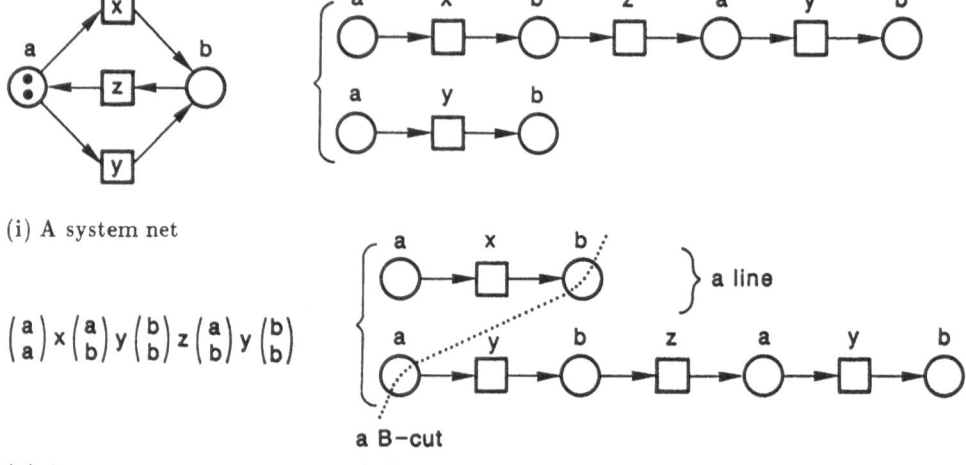

(i) A system net

$$\binom{a}{a} \times \binom{a}{b} y \binom{b}{b} z \binom{a}{b} y \binom{b}{b}$$

(ii) An occurrence sequence σ of (i)

(iii) Two occurrence nets describing processes of (i)

Fig. 3.1. A simple example

Figure 3.1 gives a simple example illustrating the main notions introduced so far (as well as anticipating the process notion).

The places a and b in Figure 3.1(i) might represent the free state and the used state, respectively, of a stock of resources. x and y then represent two (possibly conflicting) acts of claiming a resource for use, while z describes the freeing of a resource. The initial marking indicates that initially two resources are free and none is used. The occurrence sequence in Figure 3.1(ii) and the processes in Figure 3.1(iii) describe the (concurrent) claiming of the two resources and the freeing and re-claiming of one of them.

The first marking of σ in Figure 3.1(ii) enables both x and y in such a way that even if one of them occurs, the other remains enabled. We will capture this situation by the next definition.

Definition 3.2.4 [Concurrent enabling of transitions]
Let $\Sigma = (S, T, F, M_0)$ be a system net, $M \in [M_0\rangle$ a reachable marking and

t_1, t_2 two transitions. Then t_1 and t_2 are concurrently enabled by M *iff*
$$\forall s \in S \colon F(s,t_1) + F(s,t_2) \leq M(s). \qquad\qquad \Box\ 3.2.4$$

For instance, the initial marking of the net shown in Figure 3.1(i) concurrently enables x and y. We do not require t_1 and t_2 to be different, so that a transition may be concurrently enabled to itself (an example being the transition z at the third marking of σ in Figure 3.1(ii)).

We end this section by introducing the notion of safeness (often also called boundedness). Safeness means that a bound can be given for the number of tokens on a given place.

Definition 3.2.5 [Safeness]
Let $\Sigma = (S, T, F, M_0)$ be a system net and let $s \in S$.

(i) s is n-safe (for $n \in \mathbf{N}$) *iff* $\forall M \in [\, M_0 \rangle \colon M(s) \leq n$;

(ii) Σ is n-safe *iff* $\forall s \in S \colon s$ is n-safe;

(iii) Σ is safe *iff* $\exists n \colon \Sigma$ is n-safe. $\Box\ 3.2.5$

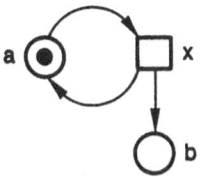

Fig. 3.2. A non-safe net

Note that all s may be n-safe (where n varies with s) without Σ itself being safe. This may only be the case for infinite nets (see Exercise 2 below). Figure 3.2 shows a net which is not safe. Clearly, a finite system net is safe iff all of its places are n-safe for some n. But a stronger characterisation of safeness can be given for finite system nets:

Theorem 3.2.6 [Characterisation of safeness for finite systems]
Let $\Sigma = (S, T, F, M_0)$ be a *finite system net*. Then the following statements are equivalent:

(i) Σ *is safe*.

(ii) $\forall M \in [\, M_0 \rangle\ \forall \tilde{M} \in [\, M \rangle \colon\ \neg (M < \tilde{M})$
       (for the meaning of $M < \tilde{M}$, see the appendix on terminology).

(iii) $[\, M_0 \rangle$ *is a finite set*.

Proof: (i) $\Rightarrow$ (ii): Assume $M < \tilde{M}$ for some $M \in [M_0\rangle$, $\tilde{M} \in [M\rangle$. Then $\exists s_0 \in S: M(s_0) < \tilde{M}(s_0)$ and $\forall s \in S: M(s) \leq \tilde{M}(s)$. Put $M_1 = M$, $M_2 = \tilde{M}$ and let τ be a transition sequence leading from M_1 to M_2. $M_2 > M_1$ implies that there is a marking M_3 such that τ leads from M_2 to M_3; of course, $M_3 > M_2$ and for the place s_0, $M_3(s_0) > M_2(s_0)$. Since τ can be reiterated arbitrarily often, s_0 cannot be a safe place.

(ii) $\Rightarrow$ (iii): We assume $[M_0\rangle$ to be infinite and we prove that there are $M \in [M_0\rangle$ and $\tilde{M} \in [M\rangle$ such that $M < \tilde{M}$. First we will construct an infinite occurrence sequence $\sigma = M_0 t_1 M_1 t_1 \ldots$ such that the M_i are mutually distinct, i.e., $i \neq j \Rightarrow M_i \neq M_j$, inductively as follows:

At step $i = 0$, choose $\sigma = M_0$.

Suppose σ to be constructed up to step $i \geq 0$, that is, $\sigma = M_0 t_1 \ldots t_i M_i$. Suppose as induction hypothesis that the set of markings reachable from M_i by occurrence sequences not containing any of the markings $\{M_0, \ldots, M_{i-1}\}$ is an infinite set (for $i = 0$, this hypothesis follows directly from the infinity of $[M_0\rangle$). Due to the finiteness of T, M_i enables only finitely many transitions. Suppose that whenever $M_i[t\rangle M$ then from M only finitely many markings can be reached by occurrence sequences not containing any of the markings $\{M_0, \ldots, M_i\}$; then from M_i only finitely many markings could be reached by occurrence sequences not containing any of the markings $\{M_0, \ldots, M_{i-1}\}$, contradicting the choice of M_i. Hence there must be a transition t_{i+1} and a marking M_{i+1} such that $M_i[t_{i+1}\rangle M_{i+1}$ and the set of markings reachable from M_{i+1} by occurrence sequences not containing any of the markings $\{M_0, \ldots, M_i\}$ is an infinite set; any such transition is suitable to prolong σ.

Let, for any fixed finite index set I, $f_0, f_1, f_2, \ldots$ be any infinite sequence of mutually distinct vectors of nonnegative integers $f_i: I \rightarrow \mathbf{N}$ $(i \geq 0)$.

Claim: There exist indices $i_0, i_1, i_2, \ldots$ with $0 \leq i_0 < i_1 < i_2 < \ldots$ and $f_{i_0} < f_{i_1} < f_{i_2} < \ldots$, i.e., an increasing subsequence of $f_0, f_1, f_2, \ldots$

Proof: By induction on the number $m = |I|$.

$m = 1$:   Then the f_i are nonnegative integers. Put $i_0 = 0$. With i_j defined, there are only finitely many integers between 0 and f_{i_j}, while the numbers f_i $(i \geq i_j)$ are mutually distinct. Hence there is an index i_{j+1} with $f_{i_{j+1}} > f_{i_j}$.

$m - 1 \rightarrow m$:   Then $f_i = (f_i', f_i'')$, where f_i' is a vector of dimension $m - 1$ and f_i'' is a nonnegative integer. If the sequence $f_0'', f_1'', f_2'', \ldots$ contains infinitely many equal elements $f_{j_0}'' = f_{j_1}'' = f_{j_2}'' = \ldots$ then we consider the infinite subsequence $f_{j_0}', f_{j_1}', f_{j_2}', \ldots$ and apply the induction hypothesis, obtaining an increasing subsequence of $f_0', f_1', f_2', \ldots$, and hence also of $f_0, f_1, f_2, \ldots$ Otherwise, $f_0'', f_1'', f_2'', \ldots$ has an infinite subsequence of mutually distinct elements. Thus, an increasing sequence $f_{j_0}'' < f_{j_1}'' < f_{j_2}'' < \ldots$ can be found by induction hypothesis. The sequence $f_{j_0}', f_{j_1}', f_{j_2}', \ldots$ contains either infinitely many equal elements $f_{i_0}' = f_{i_1}' = f_{i_2}' = \ldots$ or a subsequence of infinitely many mutually distinct elements, out of which an increasing subsequence $f_{i_0}' < f_{i_1}' < f_{i_2}' < \ldots$

can be selected by the induction hypothesis. In both cases, $f_{i_0} < f_{i_1} < f_{i_2} < \cdots$ is an increasing subsequence of $f_0, f_1, f_2, \ldots$.   □ *Claim*

Because S is finite, the sequence $\sigma_M = M_0, M_1, M_2, \ldots$ satisfies the premises of the claim, and we may find indices j, k with $M_k > M_j$ and $M_k \in [M_j\rangle$.

(iii) $\Rightarrow$ (i): If $[M_0\rangle$ is finite, $n = \max\{M(s) \mid M \in [M_0\rangle \wedge s \in S\}$ is well-defined and Σ is n-safe.   □ 3.2.6

3.3   Occurrence Nets and Processes

One may view the concurrent behaviour of a marked net again as a net with some special properties. The latter is called an occurrence net.

Definition 3.3.1 [Occurrence nets]
A net (S, T, F) is called an occurrence net *iff*

(i) $\forall s \in S: |{}^\bullet s| \leq 1 \wedge |s^\bullet| \leq 1$ and

(ii) F^+ is acyclic, i.e., $\forall x \in S \cup T: (x, y) \in F^+ \Rightarrow (y, x) \notin F^+$.   □ 3.3.1

The S-elements of an occurrence net will be called conditions and denoted by B (for German: 'Bedingung'). Conditions will be used to represent state holdings. The T-elements of an occurrence net will be called events and denoted by E. They will be used to represent transition occurrences. Definition 3.3.1(i) means that an occurrence net contains no non-deterministic choices, the idea being that all choices are resolved at the behaviour level. Definition 3.3.1(ii) means that an occurrence net contains no cycles, the idea being that all loops are unfolded at the behaviour level.

Because of Definition 3.3.1(ii), the structure $(X, \prec)$ derived from an occurrence net (B, E, F) by putting $X = B \cup E$ and $\prec = F^+$ is a partially ordered set. It is even an occurrence poset as has been defined in Section 2.5. Conversely, every occurrence poset is a poset derived from some occurrence net. With this association (which will henceforth be assumed) all the definitions and results of Chapter 2, particularly those of Sections 2.2 and 2.5, may be transferred to occurrence nets. We may, for example, speak of the cuts $C = C(N)$, the lines $L = L(N)$, etc., of an occurrence net N. For occurrence nets, we have $\prec \cdot = F$, and hence Notation 3.1.2 is consistent with Notation 2.2.1 used to denote immediate neighbours. Figure 3.3 depicts three examples of occurrence nets and their associated posets. Recall also that Figure 1.4 of the introduction shows an occurrence net.

The cuts which consist of elements of B only have a special significance. We therefore give them a special name.

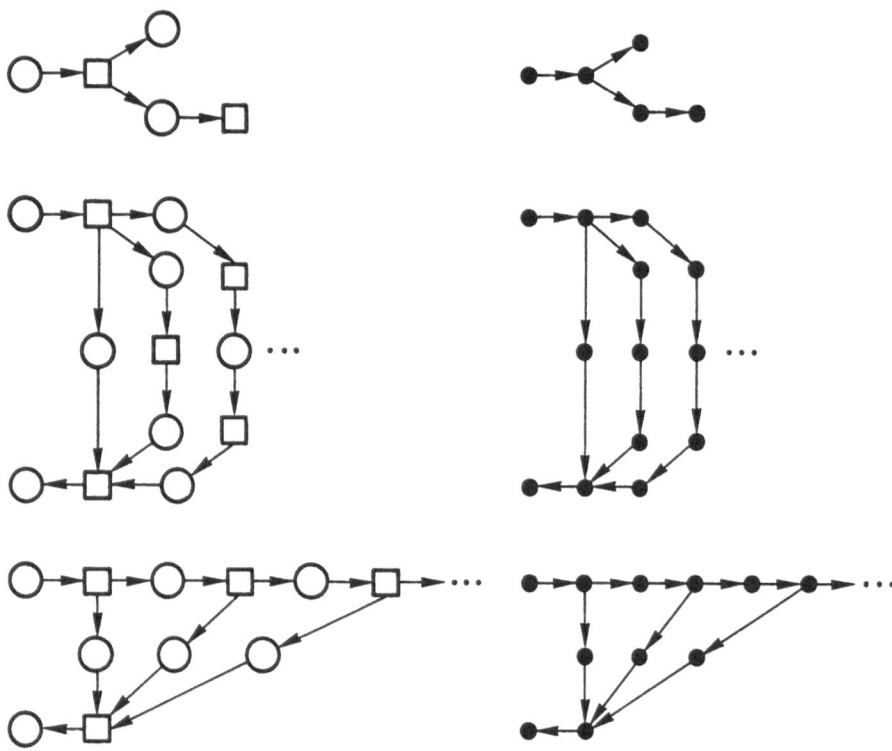

Fig. 3.3. Sample occurrence nets and their associated posets

Definition 3.3.2 [B-cuts]
Let $N = (B, E, F)$ be an occurrence net and $c \in C(N)$.
Then c is called a B-cut of N *iff* $c \subseteq B$.   □ 3.3.2

   Intuitively, the processes shown in Figure 3.1(iii) describe two of the possible concurrent behaviours of the system shown in Figure 3.1(i). There are two differences to the occurrence sequence σ of Figure 3.1(ii). Firstly, the elements of the processes are partially (rather than totally) ordered, allowing for the representation of concurrency. Secondly, only the local (rather than the global) states are represented explicitly in the processes. The markings of σ correspond to certain B-cuts of the processes.

   We shall now define processes formally. Mathematically speaking, they are labelled occurrence nets where the labelling is a function which specifies a relationship between the conditions and the events of the occurrence net and the places, the markings and the transitions of the given system net.

Each event of the occurrence net denotes a single occurrence of some transition of the system net while each condition of the occurrence net denotes the presence of a single token on some place of the system net. Thus, we are led to consider occurrence nets $N = (B, E, F')$ together with a function $p: B \cup E \rightarrow S \cup T$ linking N to a given system net $\Sigma = (S, T, F, M_0)$. The question arises: which pairs (N, p) are processes of Σ and which ones are not? In this section we will define a set of properties to answer this question. The motivation for these properties will, at first, be given intuitively. It will be strengthened in the next section where a relation to the occurrence sequences defined in Definition 3.2.2 will be proved. Let us first notice that as a consequence of our convention to require $E \subseteq dom(F') \cap cod(F')$, we have $Min(N) \subseteq B$ and $Max(N) \subseteq B$.

Definition 3.3.3 [Processes]

Let $\Sigma = (S, T, F, M_0)$ be a system net, $N = (B, E, F')$ an occurrence net and $p: B \cup E \rightarrow S \cup T$ a labelling function of N.
The pair $\pi = (N, p)$ is called a process of Σ *iff*

(i) $Min(N)$ is a B-cut of N.

(ii) N is discrete with respect to $Min(N)$ (see Definition 2.2.11).

(iii) $p(B) \subseteq S$ and $p(E) \subseteq T$.

(iv) $\forall e \in E \; \forall s \in S: F(s, p(e)) = |p^{-1}(s) \cap {}^\bullet e|$ and $F(p(e), s) = |p^{-1}(s) \cap e^\bullet|$.

(v) $\forall s \in S: M_0(s) = |p^{-1}(s) \cap Min(N)|$.   □ 3.3.3

It may be noticed that Properties 3.3.3(i) and (ii) only concern the occurrence net N, while the other properties concern the labelling function p (and hence Σ and N).

Property 3.3.3(iii) indicates that the conditions of N represent holdings of the places of Σ and that the events of N represent occurrences of the transitions of Σ. Property 3.3.3(i) means that N starts with an initial B-cut, and Property 3.3.3(v) specifies that this initial cut corresponds to the initial marking of Σ in the sense that every condition in $Min(N)$ represents some token of M_0. The meaning of Property 3.3.3(ii) has been studied extensively in Section 2.2 of this book. In Section 3.5 below, we will show that this property allows a correspondence between processes and occurrence sequences to be established. Finally, Property 3.3.3(iv) expresses the local conformity of N and Σ with respect to p, in the sense that transition environments are respected. This enshrines the transition rule defined in Definition 3.2.1(ii): the set ${}^\bullet e$ ($e^\bullet$) models the tokens consumed (produced, respectively), by the occurrence of the transition $p(e)$. The properties are independent of each other, except for $p(E) \subseteq T$ which is a necessary precondition for (iv).

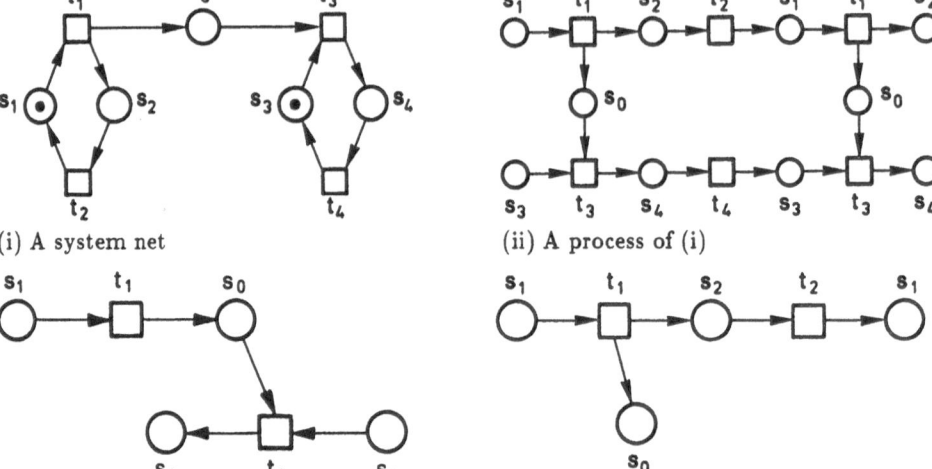

(i) A system net

(ii) A process of (i)

(iii) Not a process: Satisfies all process properties except (iv)

(iv) Not a process: Satisfies all process properties except (v)

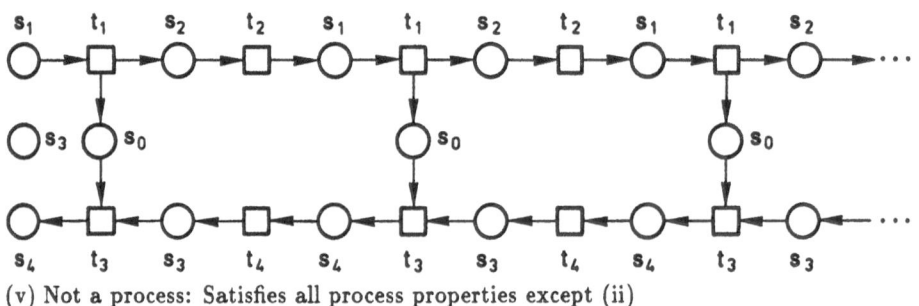

(v) Not a process: Satisfies all process properties except (ii)

Fig. 3.4. Illustration of the process definition

Figure 3.4 illustrates the process definition by means of an example.

We end this section with a few simple observations and definitions relating to the process properties. The first observation states that even though they have been stated for infinite system nets, we do not need to worry about any uncountability problems.

Theorem 3.3.4 [Processes of countable systems are countable]
Let Σ be countable and (N,p) satisfy the process properties with respect to Σ. Then $N = (B, E, F')$ is countable.

Proof: Define sets A_i as follows:

$$A_0 = Min(N), \quad A_{i+1} = A_i \cup A_i^\bullet$$

A_0 is countable because Σ is countable and because of Properties 3.3.3(iii), 3.3.3(v) and the definition of a marking.

For each i, $A_i^\bullet$ is countable because, due to the countability of Σ and Properties 3.3.3(iii,iv), $x^\bullet$ is countable for $x \in A_i$; hence A_{i+1} is also countable.

Due to Definition 3.3.3(i) and the fact that N is combinatorial,

$$B \cup E \ = \ \bigcup_{i=0}^{\infty} A_i;$$

hence, being the countable union of countable sets, $B \cup E$ is also countable.
$\square$ 3.3.4

Further, we may apply the results of Section 2.2 in order to express the discreteness property 3.3.3(ii) in different ways as follows:

Remark 3.3.5 [Equivalent reformulation of discreteness w.r.t. $Min(N)$]
In Definition 3.3.3, discreteness with respect to $Min(N)$ could be equivalently replaced by the property that N is observable with respect to $Min(N)$ (see Definition 2.2.14(iii) and Theorem 2.2.17).
$\square$ 3.3.5

A process (B, E, F', p) may be finite or infinite in the sense that $B \cup E$ is a finite set or an infinite set. There is also another reasonable notion of finiteness:

Definition 3.3.6 [Event-finiteness of processes]
Let $\pi = (B, E, F', p)$ be a process of $\Sigma = (S, T, F, M_0)$.
π is event-finite *iff* $|E| \in \mathbf{N}$.
$\square$ 3.3.6

Event-finiteness is important because it corresponds to the finiteness of occurrence sequences as has been defined in Definition 3.2.2(i).

There is also the notion of an initial part, or prefix, of a process which is analogous to what has been defined in Definition 3.2.2(ii) for occurrence sequences. Given a B-cut c of a process, we may define the portions below c and above c of the process:

Definition 3.3.7 [$\Downarrow(c, \pi)$ and $\Uparrow(c, \pi)$]
Let $\pi = (B, E, F', p)$ be a process of Σ and let c be a B-cut of π.

(i) $\Downarrow(c, \pi) \ = \ (B \cap \downarrow c \, , \ E \cap \downarrow c \, , \ F' \cap (\downarrow c \times \downarrow c) \, , \ p\,|_{\downarrow c})$.

(ii) $\Uparrow(c, \pi) \ = \ (B \cap \uparrow c \, , \ E \cap \uparrow c \, , \ F' \cap (\uparrow c \times \uparrow c) \, , \ p\,|_{\uparrow c})$.
$\square$ 3.3.7

Figure 3.5 illustrates this definition.

It is routine to check that $\Downarrow(c, \pi)$ is again a process of Σ. We will see later (in Section 3.5) under which conditions $\Uparrow(c, \pi)$ is a process of some system net derived from Σ.

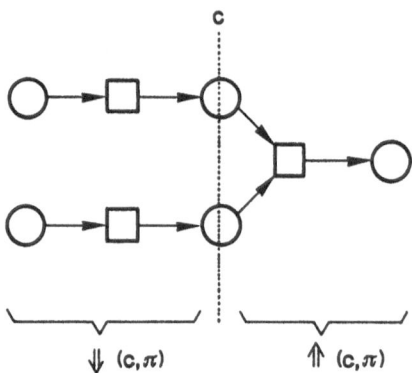

Fig. 3.5. Illustration of the definition of $\Downarrow(c, \pi)$ and $\Uparrow(c, \pi)$

When considering a process π of a system Σ, we are not interested in the exact nature of the sets used to define the conditions and the events of π. Rather, we are only interested in their interconnections and in the labelling which relates π to Σ. Hence processes which are isomorphic in the following sense shall not be distinguished because they represent the same interconnection and labelling patterns:

Definition 3.3.8 [Isomorphism of processes]
Let $\pi_1 = (B_1, E_1, F_1, p_1)$ and $\pi_2 = (B_2, E_2, F_2, p_2)$ be two processes of Σ. π_1 and π_2 are isomorphic (in symbols: $\pi_1 \cong \pi_2$) *iff* there is a bijection $\beta: B_1 \cup E_1 \rightarrow B_2 \cup E_2$ such that:

(i) $\forall x \in B_1 \cup E_1: p_1(x) = p_2(\beta(x))$ (β is label-preserving) and

(ii) $\forall x_1, x_2 \in B_1 \cup E_1: x_1 \prec_1 x_2 \iff \beta(x_1) \prec_2 \beta(x_2)$
      (β is order-preserving in both directions). $\qquad\qquad \Box$ 3.3.8

Clearly, $\cong$ is an equivalence relation. From now on, whenever we speak of two different processes we mean two non-isomorphic processes. Notice that from Definition 3.3.8(i), together with Property 3.3.3(ii), it follows that β is sort-preserving as well, that is, $\beta(B_1) = B_2$ and $\beta(E_1) = E_2$.

3.4   Inductive Definition of Processes

In the last section, the notion of a process was introduced axiomatically. Since processes are intended to capture the behaviour of a system, it should be possible to define them inductively. One may think of a process being generated by applying the transition rule several times in succession. For example, the processes shown in Figure 3.1(iii) may be related to the occurrence sequence shown in Figure 3.1(ii) in the sense that both may be generated from that sequence. The next definition formalises this idea.

Definition 3.4.1 [Associating processes to occurrence sequences]
Let $\Sigma = (S, T, F, M_0)$ be a system net and let $\sigma = M_0 t_1 M_1 \ldots$ be an occurrence sequence of Σ. To σ we associate a set $\Pi(\sigma)$ of (what will turn out to be) processes of Σ by defining a construction to produce the elements $\pi = (N, p)$ of $\Pi(\sigma)$. To this end, we construct successively embedded labelled occurrence nets $(N_i, p_i) = (B_i, E_i, F_i, p_i)$ where $p_i \colon B_i \cup E_i \to S \cup T$, by induction on i.

$i = 0$:   Define $E_0 = F_0 = \emptyset$ and B_0 as containing, for each $s \in S$, $M_0(s)$ distinct conditions b with $p_0(b) = s$ (this defines p_0 as well).

$i \to i + 1$:   Suppose that $(N_i, p_i) = (B_i, E_i, F_i, p_i)$ has already been constructed. For each $s \in {}^\bullet t_{i+1}$ we choose a condition $b(s) \in Max(N_i) \cap p_i^{-1}(s)$; then we add a new event e with $p_{i+1}(e) = t_{i+1}$ and $(b(s), e) \in F_{i+1}$ for all $s \in {}^\bullet t_{i+1}$. Also, for each $s \in t_{i+1}^\bullet$ we add a new condition $b'(s)$ with $p_{i+1}(b'(s)) = s$ and $(e, b'(s)) \in F_{i+1}$. For $x, y \in B_i \cup E_i$, we define

$$p_{i+1}(x) = p_i(x)$$
$$\text{and} \quad (x, y) \in F_{i+1} \iff (x, y) \in F_i.$$

If $\sigma = M_0 t_1 \ldots t_n M_n$ is finite, then the construction stops at $i = n$, and we put $\pi = (N, p)$ with $N = N_n$ and $p = p_n$. If σ is infinite then we put $\pi = (N, p)$ with $N = (\bigcup_i B_i, \bigcup_i E_i, \bigcup_i F_i)$ and $p = \bigcup_i p_i$.

Finally, $\Pi(\sigma) = \{\pi \mid \pi \text{ may be constructed as above }\}$. $\square$ 3.4.1

In order to show that this construction makes sense, we have to show that

$$\forall i \geq 0 \ \forall s \in {}^\bullet t_{i+1} \colon \exists b(s) \in (Max(N_i) \cap p_i^{-1}(s)).$$

This results from the fact that M_i enables t_{i+1} and from the following:

Proposition 3.4.2 [Correctness of Definition 3.4.1]
With the notation of Definition 3.4.1, we have

$$\forall s \in S \colon M_i(s) = |Max(N_i) \cap p_i^{-1}(s)|.$$

Proof: By induction on $i \geq 0$.

$i = 0$:   Immediate (notice that $Min(N_0) = B_0 = Max(N_0)$ and that B_0 is a B-cut of N_0 corresponding to M_0).

$i \to i + 1$:   By the construction, and with the notation used there,

$$Max(N_{i+1}) = (Max(N_i)\backslash\{b(s) \mid s \in {}^\bullet t_{i+1}\}) \cup \{b'(s) \mid s \in t_{i+1}^\bullet\}.$$

It is routine to check that $Max(N_{i+1})$ is a maximal co-set, i.e., a B-cut (see Exercise 6). By the transition rule 3.2.1(ii),

$$\forall s \in S\colon M_{i+1}(s) = M_i(s) - F(s, t_{i+1}) + F(t_{i+1}, s).$$

Hence the property.   □ 3.4.2

The construction used in Definition 3.4.1 is non-deterministic because, in general, there may be more than one $b(s) \in Max(N_i) \cap p_i^{-1}(s)$. For example, both processes shown in Figure 3.1(iii) are in $\Pi(\sigma)$ if σ is the occurrence sequence shown in Figure 3.1(ii).

The fact that Definition 3.4.1 yields only processes will be expressed as the main theorem of this section:

Theorem 3.4.3 [The inductive Definition 3.4.1 yields processes]
All $\pi \in \Pi(\sigma)$ satisfy the process properties 3.3.3.

Proof: It is routine to check the properties 3.3.3 for $\pi = (N, p)$ as well as for each (N_i, p_i) used in Construction 3.4.1:

$Min(N) = Min(N_i)$ (all $i \geq 0$) is a B-cut by construction, so that Property 3.3.3(i) holds.

If $e \in E$ has been constructed at step $i \geq 1$ of Construction 3.4.1 then for all $l \in L(N_i)$:

$$|l \cap [Min(N_i), e]| \leq 2 * i,$$

and this bound is also valid for all $l \in L(N)$ since events and conditions constructed at step $j > i$ cannot be contained in the set $l \cap [Min(N_j), e]$, for any $l \in L(N_j)$; hence Property 3.3.3(ii) holds.

Property 3.3.3(iii) holds true because

$$\begin{aligned} p(B) &= \textstyle\bigcup_{i=0}^\infty p(B_i) \subseteq S \\ \text{and} \quad p(E) &= \textstyle\bigcup_{i=0}^\infty p(E_i) \subseteq T \end{aligned}$$

by construction.

The local conformity property 3.3.3(iv) and the initial marking conformity property 3.3.3(v) result immediately from the construction.   □ 3.4.3

The preceding result can be interpreted in two ways. If one accepts the axiomatic definition of a process via Definition 3.3.3 then one may be assured that the inductive Definition 3.4.1 does not produce anything which fails to be a process. If one accepts the inductive Definition 3.4.1, then the theorem states that Properties 3.3.3 do not exclude some of the objects definable by Definition 3.4.1. In both cases, the converse question arises: Do there exist processes in the sense of Definition 3.3.3 which cannot be produced inductively through Construction 3.4.1? We shall turn to this question in the next section. Before answering it, however, we shall use the remainder of this section to introduce some notations and elementary results relating to the inductive definition which will be needed later. For the rest of the section we fix a system net $\Sigma = (S, T, F, M_0)$.

Definition 3.4.4 [The Lin-set of a process]
Let π be a process of Σ.
$Lin(\pi) = \{\sigma \mid \sigma$ is an occurrence sequence of Σ and $\pi \in \Pi(\sigma)\}$.   □ 3.4.4

In a sense, the Lin operator is the inverse of Π. However, notice that $Lin(\pi)$ does not comprise all linear extensions of π. Rather, $Lin(\pi)$ contains only those linear extensions of π that are at the same time occurrence sequences.

Definition 3.4.5 [Position function]
Let $\sigma = M_0 t_1 M_1 t_2 \ldots$ be an occurrence sequence of Σ and let $\pi = (B, E, F', p)$ be a process which has resulted from an application of Construction 3.4.1 to σ. Then we denote by $pos_\sigma(e)$, for $e \in E$, the number i of the step at which $e = e_i$ has been constructed.   □ 3.4.5

For an event e of π, the function pos_σ yields the index of the corresponding transition in σ. This index depends on σ, on π, as well as on the particular construction by which π has been derived from σ. The important facts are, however, that a function pos_σ can be defined whenever $\pi \in \Pi(\sigma)$ and, moreover, that any such function satisfies the following properties:

Proposition 3.4.6 [Properties of the pos function]
With the notation of Definition 3.4.5,

(i) pos_σ *is a bijection from the set of events of π to the set of transition indices of σ.*

(ii) $t_{pos_\sigma(e)} = p(e)$, *for all $e \in E$ (pos_σ is label-preserving).*

(iii) $\forall e_1, e_2 \in E : e_1 \prec e_2 \Rightarrow pos_\sigma(e_1) < pos_\sigma(e_2)$ *(pos_σ is order-preserving).*

Proof: This is an immediate consequence of Construction 3.4.1.   □ 3.4.6

Sometimes, when allowed by the context, we shall omit the index σ of *pos*. Notice that Properties 3.4.6(i) and (iii) imply that pos_σ is an injective observer of the poset $(E, \prec |_{E \times E})$.

The next proposition gives a correspondence between the sequence of markings of σ and a sequence of B-cuts of an associated π.

Proposition 3.4.7 [Retrieving markings as B-cuts]
Let $\sigma = M_0 t_1 M_1 t_2 \dots$ *be an occurrence sequence of* Σ *and let* $\pi = (N, p) = (B, E, F', p)$ *be any process in* $\Pi(\sigma)$, *with a position function* pos_σ. *Then there are B-cuts* $c_0, c_1, \dots$ *in* π *such that for all* $i \geq 0$:

(i) $c_i \sqsubseteq c_{i+1}$ (see Exercise 2 of Chapter 2).

(ii) $\forall s \in S\colon M_i(s) = |c_i \cap p^{-1}(s)|$.

(iii) $c_{i+1} = (c_i \setminus {}^\bullet pos_\sigma^{-1}(i+1)) \cup pos_\sigma^{-1}(i+1)^\bullet$.

Proof: By Definition 3.4.1 and Proposition 3.4.2, if $c_i = Max(N_i)$ (where N_i are the approximations of N used in Definition 3.4.1), Properties (i)-(iii) are satisfied. It remains to be shown that the $c_i = Max(N_i)$ are B-cuts in N (as well as in N_i). First, it is clear that c_i is a co-set in N. To show the co-maximality of c_i, consider any element $x \in B \cup E$. If $\exists j \leq i\colon x \in B_j \cup E_j$ then $x \preceq y$ for some $y \in c_i$ because c_i is a B-cut in N_i. Otherwise, $x \in B_j \cup E_j$ for some $j > i$ and, by construction, we have that, successively, $x \succeq x_1$ for some $x_1 \in c_{j-1}$, $x_1 \succeq x_2$ for some $x_2 \in c_{j-2}$, ..., $x_{j-i} \succeq y$ for some $y \in c_i$.   □ 3.4.7

There may be several sequences c_i with Properties (i) and (ii) given in Proposition 3.4.7; however, because of Property (iii) the sequence in question is uniquely defined by $c_0 = Min(N)$ and pos_σ. We capture this in the following definition.

Definition 3.4.8 [Correspondence between markings and B-cuts]
Let $\sigma = M_0 t_1 M_1 t_2 \dots$ be an occurrence sequence of Σ, $\pi = (B, E, F', p)$ a process in $\Pi(\sigma)$ and pos_σ an associated position function. Then a marking M_i of σ and a B-cut c of π will be said to correspond to each other *iff* c is the unique c_i of Proposition 3.4.7.   □ 3.4.8

Finally, we will investigate Construction 3.4.1 for the important case of 1-safe nets.

Theorem 3.4.9 [The special case of 1-safe nets]
Let Σ *be 1-safe and let* $\sigma = M_0 t_1 M_1 t_2 \dots$ *be an occurrence sequence of* Σ. *Then all processes in* $\Pi(\sigma)$ *are isomorphic to each other.*

Proof: The construction in Definition 3.4.1 is in this case deterministic (up to the choice of nodes), since $M_i(s) \leq 1$ (by 1-safeness) implies

$$|Max(N_i) \cap p_i^{-1}(s)| \leq 1$$

(by Proposition 3.4.2). Hence the process π that may be constructed from σ is unique up to isomorphism.   □ 3.4.9

When isomorphic processes are identified then the statement of Theorem 3.4.9 may be expressed more simply as $|\Pi(\sigma)| = 1$. In other words, for 1-safe nets, every occurrence sequence has one and only one associated process.

3.5   Systems of Finite Synchronisation

We turn to the question left open in connection with Theorem 3.4.3: can all objects satisfying the process definition 3.3.3 be produced by means of Construction 3.4.1? Briefly, the answer will be negative in general, but a wide class of nets can be identified for which Definitions 3.3.3 and 3.4.1 indeed define exactly the same objects. These nets will be called of finite synchronisation. Finite synchronisation is related to the property of finite degree introduced in Definition 2.2.8.

First we turn to a counterexample. Figure 3.6 shows an infinite system net (Figure 3.6(i)) and a labelled occurrence net (Figure 3.6(ii)) which satisfies all of the process properties 3.3.3. There is clearly no occurrence sequence which can generate the process shown in Figure 3.6(ii); for example, an attempt $x x_1 x_2 \ldots$ (markings omitted) will fail to include the transition x'. On the other hand, we may argue that the process shown in Figure 3.6(ii) does describe a meaningful behaviour of the system shown in Figure 3.6(i). This example shows that, in general, processes in the sense of Definition 3.3.3 describe a wider class of behaviours than do occurrence sequences and the processes derived from them by means of Construction 3.4.1. However, there is a large class of system nets for which a converse of Theorem 3.4.3 holds true:

Definition 3.5.1 [Nets of finite synchronisation]
A net (S, T, F) is of finite synchronisation *iff* $\forall t \in T: |{}^\bullet t| \in \mathbf{N} \wedge |t^\bullet| \in \mathbf{N}$.
                                                                        □ 3.5.1

Finite synchronisation is related to, but not to be confused with, a property known in computer science by the name of bounded nondeterminism. Figure 3.7 explains the difference.

Bounded nondeterminism postulates that the number of nondeterministic choices in any given state is finite. Thus the structure $|s^\bullet| \notin \mathbf{N}$ shown in Figure 3.7(i) is excluded. By contrast, finite synchronisation postulates that there is no synchronisation between infinitely many participants, thus excluding the structure $|{}^\bullet t| \notin \mathbf{N}$ shown in Figure 3.7(ii).

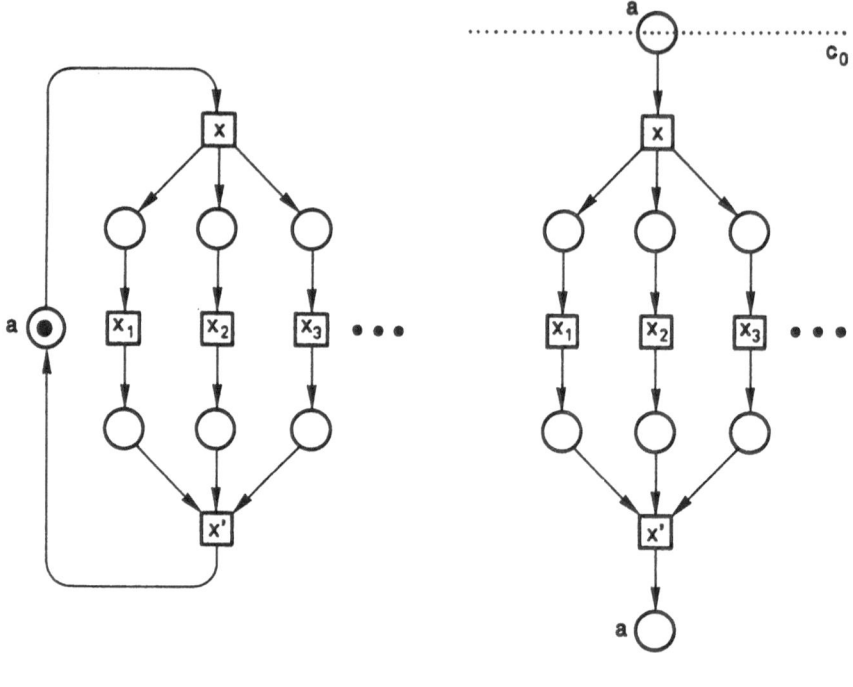

(i) A system net   (ii) A process of (i)

Fig. 3.6. A process which cannot be generated by occurrence sequences

(i) Unbounded nondeterminism   (ii) Infinite synchronisation

Fig. 3.7. Bounded nondeterminism versus finite synchronisation

Remark 3.5.2 [Finite synchronisation and degree-finiteness]
If Σ is of finite synchronisation then all processes of Σ are of finite degree.

$\square$ 3.5.2

Notice that the net shown in Figure 3.6(i) is not of finite synchronisation and that the occurrence net shown in Figure 3.6(ii) is not of finite degree.

Using this concept, the main result of this section can be stated as follows:

Theorem 3.5.3 [Equality of axiomatic and inductive process definitions]
Let $\Sigma = (S, T, F, M_0)$ be a system net of finite synchronisation and let $\pi = (N, p) = (B, E, F', p)$ be a process of Σ according to Definition 3.3.3. Then there exists an occurrence sequence σ of Σ such that $\pi \in \Pi(\sigma)$.

Proof: First suppose that $\pi = (N, p) = (B, E, F', p)$ is event-finite, i.e., $|E| \in \mathbf{N}$. Since E is finite and since $F'^+ = \prec$ is a partial ordering on E, it is possible to number the elements of E in such a manner that $E = \{e_1, e_2, \ldots, e_{|E|}\}$ and $e_j \prec e_i$ in N implies $j < i$ (i.e., the numbering is a total ordering compatible with the given partial ordering). Given such a numbering we define $t_i = p(e_i)$ ($t_i \in T$ by Property 3.3.3(iii)) and sets $c_0, \ldots, c_{|E|} \subseteq B$ inductively as follows (where pre-sets and post-sets are taken with respect to F'):

$$c_0 = Min(N), \quad c_i = (c_{i-1} \setminus {}^\bullet e_i) \cup e_i^\bullet \quad (0 < i \leq |E|)$$

(the fact that $c_i \subseteq B$ results from Property 3.3.3(i) and the definition of the cuts c_i). We claim that the c_i are B-cuts, that $[Min(N), c_i] \cap E = \{e_1, \ldots, e_i\}$ and that ${}^\bullet e_i \subseteq c_{i-1}$ for $0 < i \leq |E|$. (The proof of these facts is purely routine and is relegated to Exercise 9.)

Next, if we define for $0 \leq i \leq |E|$ and for $s \in S$: $M_i(s) = |c_i \cap p^{-1}(s)|$ then it is easy to see, by induction on i, that due to Properties 3.3.3(iv,v) and the fact that ${}^\bullet e_i \subseteq c_{i-1}$, for $0 < i \leq |E|$: M_{i-1} is a marking enabling t_i and $M_{i-1}[t_i\rangle M_i$. This completes the construction of an occurrence sequence $\sigma = M_0 t_1 \ldots t_n M_n$ of Σ such that $\pi \in \Pi(\sigma)$; it is easy to see that $pos_\sigma(e_i) = i$ ($0 < i \leq |E|$) defines a position function and M_i corresponds to c_i ($i \geq 0$) in the sense of Definition 3.4.8.

It remains to consider the case that E is infinite. Because E is countable by Theorem 3.3.4, it can be numbered. We need to show that E can be numbered in the following particular way: $E = \{e_1, e_2, \ldots\}$ such that $e_j \prec e_i$ in N implies $j < i$; to show this, Property 3.3.3(ii) and the results of Section 2.2 will be used. From Property 3.3.3(ii), i.e., discreteness with respect to $Min(N)$, Theorems 2.2.15(i), 2.2.17 and the (trivial) fact that observability with respect to c implies observability in general, N is boundedly discrete and hence weakly discrete by Theorem 2.2.10(ii). Because Σ is of finite synchronisation, N is degree-finite by Remark 3.5.2. By degree-finiteness and Theorem 2.2.10(iii), N is interval-finite. By Theorem 2.2.16 and the countability of N, there exists an injective observer of N; clearly, any injective observer yields, by definition, the desired numbering of the set E. Given a numbering of E which is compatible with the original partial ordering, an infinite occurrence sequence $\sigma = M_0 t_1 M_1 t_1 \ldots$ with $\pi \in \Pi(\sigma)$ can be defined inductively in exactly the same way as before; we omit the details because they are the same ones as if E is finite.   □ 3.5.3

Remark 3.5.4

The essential consequence of degree-finiteness, together with Property 3.3.3(ii), is the fact that E can be linearised in an appropriate way, i.e., such that the resulting linear order is both compatible with the given partial order and order-isomorphic to (a subset of) the natural numbers. The remaining part of the proof of Theorem 3.5.3 consists of the standard construction of suitable B-cuts.

This construction works for all possible appropriate linearisations of E, that is, for all injective observers on the set E, yielding sequences of pairwise corresponding B-cuts and markings. If $e \ co \ e'$ in π then e and e' can be linearised either way, so that there exist two sequences σ_1, σ_2 in $Lin(\pi)$ and position functions such that $pos_{\sigma_1}(e) < pos_{\sigma_1}(e')$ and $pos_{\sigma_2}(e') < pos_{\sigma_2}(e)$.

The construction also works if only half of the property of finite synchronisation – either $\forall t \in T \colon |{}^{\bullet}t| \in \mathbf{N}$ or $\forall t \in T \colon |t^{\bullet}| \in \mathbf{N}$ – is assumed.   □ 3.5.4

Remark 3.5.5

Property 3.3.3(ii), together with finite synchronisation, is only a sufficient (not a necessary) condition for the possibility of linearising E suitably. The exact condition is (with the notation used in Definition 3.3.3):

$$\forall e \in E \colon |E \cap [Min(N), e]| \in \mathbf{N}$$

(see Exercise 10 below).   □ 3.5.5

We shall end this section by stating a set of auxiliary results intended to deepen the understanding of the relationship between processes and occurrence sequences. These results will be needed in Chapter 4. If π is a process of Σ then there may well be B-cuts of π which do not correspond to any marking in any $\sigma \in Lin(\pi)$. An example is furnished by Figure 3.8.

The reachable B-cuts of a system net of finite synchronisation are characterised by the following definition:

Definition 3.5.6 [Reachable B-cut]

Let Σ be a system net of finite synchronisation, $\pi = (N, p) = (B, E, F', p)$ a process of Σ and c a B-cut of N.
c is reachable *iff* $|E \cap [Min(N), c]| \in \mathbf{N}$.   □ 3.5.6

This terminology is justified by Theorem 3.5.7:

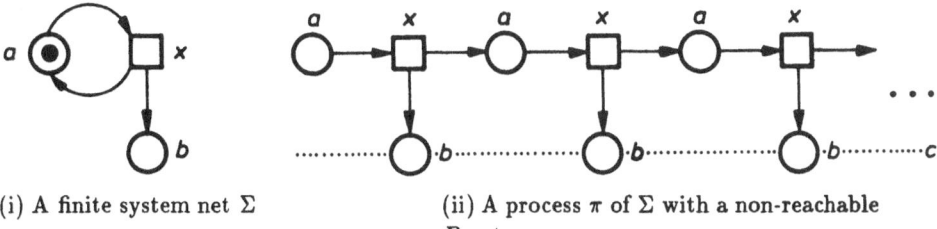

(i) A finite system net Σ (ii) A process π of Σ with a non-reachable
                                                   B-cut c

Fig. 3.8. An example of a non-reachable B-cut

Theorem 3.5.7 [Reachable B-cuts correspond to reachable markings]
Let $\Sigma = (S, T, F, M_0)$ be a system net of finite synchronisation, $\pi = (N, p) = (B, E, F', p)$ a process of Σ and c a reachable B-cut of Σ.
Then there is a marking $M \in [\, M_0\rangle$ and an occurrence sequence leading from M_0 to M such that M corresponds to c in the sense of Definition 3.4.8.

Proof: First, note that $\Downarrow(c, \pi)$ (see Definition 3.3.7) is again a process of Σ and that $c = Max(\Downarrow(c, \pi))$. Using the proof of Theorem 3.5.3, a sequence of cuts $c_0, c_1, \ldots$ and a sequence of events $e_1, e_2, \ldots$ may be constructed which satisfy $c_0 = Min(\Downarrow(c, \pi))$ and $c_i = (c_{i-1} \setminus {}^\bullet e_i) \cup e_i^\bullet$ for $0 < i$. Because c is reachable, both sequences must be finite, i.e., of the form $c_0, \ldots, c_m$ and $e_1, \ldots, e_m$, respectively, where $m = |E \cap [\, Min(N), c]|$; moreover, $c = c_m$. With $t_i = p(e_i)$ ($0 < i \leq m$) and $\forall s \in S \colon M_i(s) = |c_i \cap p^{-1}(s)|$ ($0 \leq i \leq m$), $M_0 t_1 \ldots t_m M_m$ is an occurrence sequence leading to a marking $M = M_m$ that corresponds to c_m in the sense of Definition 3.4.8.   □ 3.5.7

As a consequence of Theorem 3.5.7, if π is a process of $\Sigma = (S, T, F, M_0)$ (Σ of finite synchronisation) and c is a reachable B-cut of π then $\Uparrow(c, \pi)$ is a process of the system net $\Sigma' = (S, T, F, M)$ where $M \in [\, M_0\rangle$ as in the proof of Theorem 3.5.7.

Finally, we show that every finite B-co-set can be reached, i.e., is contained in some reachable B-cut.

Theorem 3.5.8 [Finite B-co-sets are reachable]
Let Σ be a system net of finite synchronisation, $\pi = (N, p)$ a process of Σ and c_0 a finite B-co-set of $N = (B, E, F')$ (i.e., c_0 is finite, is a co-set, and $c_0 \subseteq B$).
Then there is a reachable B-cut c such that $c_0 \subseteq c$.

Proof: Define

$$A = Min(N) \cup [\, Min(N), c_0] \cup \{b \in B \mid {}^\bullet b \subseteq [\, Min(N), c_0]\}$$

and

$$c = Max(A) = \{b \in A \mid b^\bullet \cap A = \emptyset\}.$$

We claim that c is a reachable B-cut containing c_0.

First, $c \subseteq B$ since, due to the third component of A, $\forall e \in A \cap E: e^\bullet \subseteq A$. Moreover, $\forall b' \in c_0: \{e\} = b'^\bullet \Rightarrow e \notin A$, otherwise $e \in [Min(N), c_0]$ and c_0 would not be a co-set; consequently, $c_0 \subseteq c$. If $b', b'' \in c$ and $b' \prec b''$, we have for $\{e'\} = b'^\bullet$, $\{e''\} = {}^\bullet b''$: $e' \preceq e''$, $e'' \in [Min(N), c_0]$ and $e' \in [Min(N), c_0]$ but that contradicts the fact that $b' \in c$; thus c is a co-set.

Let $y \in (B \cup E) \setminus c$. If $y \notin A$, as $Min(N)$ is a B-cut, there is a finite directed chain from $Min(N)$ to y. Let z be the last element in A on that chain. From the previous considerations, it follows that $z \in c$ and consequently y is not co with c. If $y \in A \cap E$ then $y \in [Min(N), c_0]$ and y is not co with c_0. If $y \in (A \cap B) \setminus c$ then $y^\bullet \subseteq [Min(N), c_0]$ and again, y is not co with c_0. Consequently, c is a maximal co-set, i.e., a B-cut.

Finally, we prove that c is reachable. To this end, we first show (using the fact that Σ is of finite synchronisation and the process property 3.3.3(ii), i.e., discreteness with respect to $Min(N)$) that $[Min(N), e]$ is finite for each $e \in E$. Let $e \in E$; by discreteness with respect to $Min(N)$, a natural number $n(e) \in \mathbf{N}$ can be defined which denotes the length of a longest li-set between $Min(N)$ and e. We prove $|[Min(N), e]| \in \mathbf{N}$ by induction on $n(e)$.

If $n(e) = 2$ then ${}^\bullet e \subseteq Min(N)$ and $[Min(N), e] = {}^\bullet e \cup \{e\}$ is finite by the degree-finiteness of N.

If $n(e) > 2$ then $[Min(N), e] = \{e\} \cup {}^\bullet e \cup \bigcup_{e' \in {}^{\bullet\bullet}e} [Min(N), e']$ is finite by the degree-finiteness of N and the induction hypothesis since $e' \in {}^{\bullet\bullet} e$ implies $n(e') \leq n(e) - 2$. To prove that c is reachable we note that

$$
\begin{aligned}
|E \cap [Min(N), c]| &= |E \cap [Min(N), c_0]| \\
&\leq \textstyle\sum_{b \in c_0} |E \cap [Min(N), b]| \\
&\leq \textstyle\sum_{b \in c_0 \setminus Min(N)} |[Min(N), {}^\bullet b]|.
\end{aligned}
$$

The last sum is finite because of the finiteness of c_0 and because $[Min(N), e]$ is finite, as we have seen.   □ 3.5.8

The main result of this section, i.e., Theorem 3.5.3, can be interpreted as a converse of Theorem 3.4.3 for the case of systems of finite synchronisation. It states that for such systems, the objects described by the process definition 3.3.3 are exactly the same as the objects defined inductively from occurrence sequences via Definition 3.4.1.

This allows some discussion of the adequacy of the definitions which have been given using occurrence sequences. We have defined the set of reachable markings $[M_0\rangle$ in Definition 3.2.2(iii) using occurrence sequences. This complies with the Definition 3.5.6 of a reachable cut (see Exercise 11). By Definition 3.5.6, not all B-cuts of the process shown in Figure 3.6(ii) are reachable. The process properties given in Definition 3.3.3 suggest a different notion of reachability which would, for instance, make all B-cuts of Figure 3.6(ii) reachable (see Exercise 12). Using an argument as in Theorem 3.5.3, it can be shown that provided M_0 is a finite marking, then for systems of finite synchronisation, we do not need to worry about such a different definition (see also Exercise 13).

Theorem 3.5.3 can also be interpreted more widely as a formal basis for comparing the 'true concurrency semantics' (i.e., the processes of a net) with the 'arbitrary interleaving semantics' (i.e., the occurrence sequences of a net). One may hope to deduce from it what concepts can – or cannot – be defined equally well in terms of the two semantics, as was done above for the set of reachable markings.

Exercises.

1. Calculate $[M_0\rangle$, according to Definition 3.2.2(iii),

   (i)   for the net shown in Figure 3.1(i);

   (ii)   for the net shown in Figure 3.6(i);

   (iii) for the net shown in Figure 3.8(i).

2. Invent an (infinite) unsafe net such that each place s is n-safe (where n depends on s).

3. Which of the implications of Theorem 3.2.6 become wrong for nets in which S (T) is infinite / for nets of finite nondeterminism?

4. Prove or disprove that if the occurrence net N used in Definition 3.3.3 has finitely many events then the process definition can be simplified by omitting (ii) and (iii).

5. Prove that with the notation of Definition 3.3.7, $\Downarrow (c, \pi)$ is a process of the system net Σ.

6. Complete the proof of Proposition 3.4.2.

7. Calculate $\Pi(\sigma)$, give the *pos* functions and all correspondences between markings, B-cuts, transitions and events:

   (i)   For $\sigma = xyzxyz$ (markings omitted) of Figure 3.1(i).

   (ii)   For $\sigma = xx_1x_2x_3\ldots$ (markings omitted) of Figure 3.6(i).

   (iii) For $\sigma = xxx\ldots$ (markings omitted) of Figure 3.8(i).

8. Calculate $Lin(\pi)$ and give the *pos* function(s) as well as all correspondences between markings, B-cuts, transitions and events for the processes shown in Figure 3.1(iii).

9. Complete the proof of Theorem 3.5.3.

10. Show that if Property 3.3.3(ii) is replaced by

$$\forall e \in E \colon |E \cap [\,Min(N),e]| \in \mathbf{N},$$

then Theorem 3.4.3 remains true, while Theorem 3.5.3 holds without the premise of finite synchronisation.

11. Define $[\,\widetilde{M_0})$ as the set of markings of $\Sigma = (S,T,F,M_0)$ which corresponds to the reachable B-cuts of processes of Σ and show that $[\,M_0) = [\,\widetilde{M_0})$ (even if the premise of finite synchronisation is dropped).

12. Let $\Sigma = (S,T,F,M_0)$ be a system net, $\pi = (N,p) = (B,E,F',p)$ a process of Σ and c a cut of N. Define c to be g-reachable ('generalised-reachable') *iff*

$$\exists n \in \mathbf{N}^+ \, \forall l \in L(N) \colon |l \cap [\,Min(N),c]| < n.$$

Find a system net of finite synchronisation and a process of it which has a B-cut that is g-reachable but not reachable, and a B-cut that is not g-reachable.

13. Show:

(i)   Every reachable cut is g-reachable.

(ii)   Let π be a process of a system Σ such that Σ is of finite synchronisation and has a finite initial marking M_0 (that is, the set of places s with $M_0(s) > 0$ is finite); then every g-reachable cut of π is reachable.

14. Calculate all *non*-reachable B-cuts

(i)   in Figure 3.6(ii);

(ii)   in Figure 3.8(ii).

15. Calculate all finite B-co-sets in Figure 3.8(ii) and show that Theorem 3.5.8 is satisfied.

Chapter 4. Connections Between Systems and Processes

4.1 Introduction

Having defined the processes of a system net via the process definition 3.3.3, or (as we know: equivalently for system nets of finite synchronisation) via the construction given in Definition 3.4.1, it is now possible to translate properties from a system to the set of its processes and vice versa. It is possible to ask, for example, what it would mean for a system that all of its processes enjoy some interesting property, or conversely.

In this chapter, we shall present two results of this kind. The first establishes a relationship between K-density as investigated in Sections 2.3 and 2.5, and system safeness as defined in Definition 3.2.5. The result states that a finite system net is safe iff all of its processes are K-dense; thus, in a sense, safeness is the 'system equivalent' of K-density.

The second result establishes a similar relationship between D-continuity and another system property to be defined in this chapter. Of course, the latter must be stronger than safeness, bearing in mind the fact that according to Theorem 2.4.14, D-continuity is stronger than K-density.

In this chapter we will restrict ourselves to finite system nets, the reason being that the results would be wrong for infinite systems, even if they are of finite synchronisation. Because of this restriction, we know from the results of the previous chapter that we may use the process definitions 3.3.3 and 3.4.1 interchangeably, and that Definition 3.2.2(iii) of the set $[M_0\rangle$ of reachable markings can be used without problems; we will do so extensively in this chapter. We also recall that we require $T \subseteq dom(F) \cap cod(F)$ for every net $N = (S, T, F)$.

4.2 K-density and Safeness

Figure 4.1 shows that a non-K-dense occurrence net can be the process of a finite unsafe system.

The next theorem shows that this connection holds true in general:

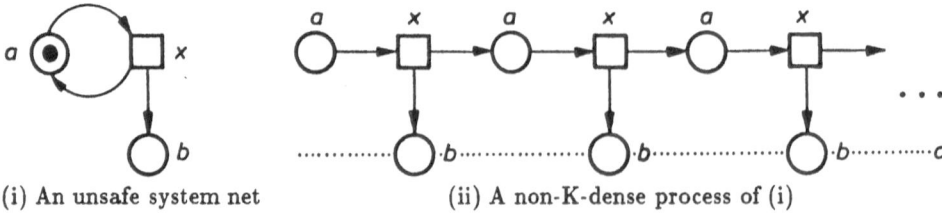

(i) An unsafe system net   (ii) A non-K-dense process of (i)

Fig. 4.1. Illustration of the relation between K-density and safeness

Theorem 4.2.1 [Relationship between K-density and safeness]
Let $\Sigma = (S, T, F, M_0)$ be a finite system net.
Then Σ is safe iff for all processes $\pi = (N, p)$ of Σ, N is K-dense.

Proof: We prove this theorem in the equivalent indirect form: 'There exists
a process $\pi = (N, p)$ of Σ such that N is not K-dense iff Σ is unsafe.'

($\Rightarrow$)   Suppose $\pi = (N, p)$ is a process of Σ and $N = (B, E, F')$ is not K-dense.
By Theorem 2.5.5(i), there is an infinite B-cut c_1 in N. Define $s_1 \in S$ such that
$|p^{-1}(s_1) \cap c_1| \notin \mathbf{N}$; such a place s_1 exists because S is a finite set. Now let $n \in \mathbf{N}$
be arbitrary. There exists a finite co-set $c_0 \subseteq c_1$ such that $|p^{-1}(s_1) \cap c_0| \geq n$.
By Theorem 3.5.8, c_0 is contained in a reachable B-cut of N (c_1 itself may
not be reachable), say $c \supseteq c_0$. By Theorem 3.5.7, there is a reachable marking
$M \in [M_0\rangle$ which corresponds to c, i.e., $\forall s \in S \colon M(s) = |p^{-1}(s) \cap c|$. In
particular, $M(s_1) \geq n$. Since n was arbitrary, Σ is not safe.

($\Leftarrow$)   Suppose that Σ is unsafe. Since Σ is finite, we may use Theorem 3.2.6
to deduce that there are markings $M_1 \in [M_0\rangle$ and $M_2 \in [M_1\rangle$ such that
$M_1 < M_2$. $M_1 < M_2$ implies that there is some place $s_1 \in S$ such that
$M_1(s_1) < M_2(s_1)$. Let τ_0 be a transition sequence leading from M_0 to M_1, let
τ_1 be a transition sequence leading from M_1 to M_2 and let the infinite sequence
τ be defined as follows:

$$\tau = \tau_0 \tau_1 \tau_1 \tau_1 \ldots$$

Because $M_2(s) \geq M_1(s)$ for all $s \in S$, τ defines a transition sequence, and
hence also an occurrence sequence σ, of Σ. σ is of the form:

$$\sigma = M_0 \sigma_0 M_1 \sigma_1 M_2 \sigma_2 \ldots$$

where σ_0 corresponds to τ_0 and σ_j corresponds to τ_1 for all $j \geq 1$, and, moreover,
$M_{j+1}(s_1) > M_j(s_1)$ for all $j \geq 1$.
      Now we may use Construction 3.4.1 to associate a suitable process $\pi = (N, p)$ to σ as follows: Let $|\tau_0| = i_0$ and $|\tau_1| = i_1$, where $|\tau|$ denotes the length
of τ, i.e., the number of transitions occurring in τ. Up to step $i_0 + i_1$, we perform
the construction arbitrarily. At each step $i_0 + (j * i_1)$, $j \geq 1$, a condition b_j with

$p(b_j) = s_1$ is set apart which never receives an output event in any further step of the construction; this is possible because $M_{j+1}(s_1) > M_j(s_1)$ and for all $s \in S$, $M_{j+1}(s) \geq M_j(s)$. In π there are conditions $b_1, b_2, \ldots$ with $p(b_j) = s_1$ ($j \geq 1$) and $b_1^\bullet = b_2^\bullet = \ldots = \emptyset$; as a consequence of the latter, $\{b_1, b_2, \ldots\}$ is a co-set. Let c be any B-cut with $c \supseteq \{b_1, b_2, \ldots\}$; then c is infinite. Because Σ is finite, $Min(N)$ is a finite B-cut, and N is of finite degree. By Theorem 2.3.12 of Chapter 2, N cannot be K-dense. □ 4.2.1

In Section 2.3, K-density had been motivated as a nice property by means of the interpretation of lines and cuts given there. The present theorem now shows that K-density indeed corresponds to a nice system property, namely safeness. Exercise 1 at the end of this chapter shows that a restricted kind of K-density holds always.

4.3   D-continuity and Frozen Tokens

In order to find a system property which corresponds to D-continuity we recall Theorem 2.5.8 which characterises D-continuity for occurrence posets (and a fortiori for occurrence nets):

$$\text{D-continuity} \iff \underbrace{\text{K} - \text{density} \wedge \text{Cut} - \text{boundedness}}_{\text{Gap-freeness}} \wedge \underbrace{\text{Non} - \text{single degree property}}_{\text{Jump-freeness}}.$$

The system equivalent of K-density is known from the results of the last section. Hence we may concentrate first on the cut-boundedness part of the characterisation of D-continuity. Figure 4.2(i) shows a system which generates the non-cut-bounded process shown in Figure 4.2(ii).

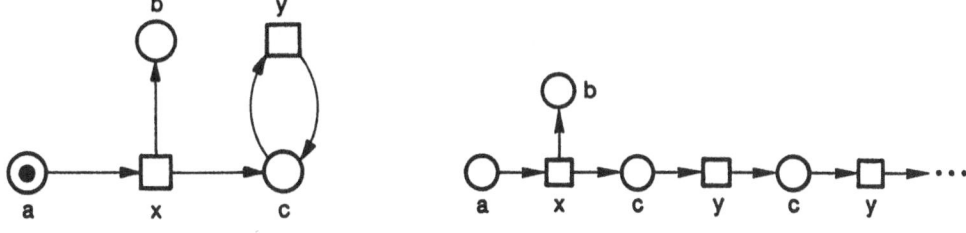

(i) A system net with a frozen token

(ii) A process of (i) which is not cut-bounded

Fig. 4.2. A system with a non-cut-bounded process

We note that in this process, there is a token on the place b which remains unmoved. We call such tokens 'frozen' and characterise them by the following definition:

Definition 4.3.1 [Frozen tokens]
A system net $\Sigma = (S, T, F, M_0)$ contains a frozen token *iff* there exist a place $s \in S$ and an infinite occurrence sequence $M_1 t_1 M_2 t_2 \ldots, M_1 \in [\, M_0\rangle$, such that

$$\forall i \geq 1: \ M_i(s) \geq \begin{cases} 1 & \text{if } s \notin {}^\bullet t_i \\ 2 & \text{if } s \in {}^\bullet t_i. \end{cases}$$

□ 4.3.1

If Σ has no side condition, i.e., no place s with $s^\bullet \cap {}^\bullet s \neq \emptyset$, then the last requirement in Definition 4.3.1 could be simplified to $\forall i \geq 1: M_i(s) \geq 1$. The clause $M_i(s) \geq 2$ takes effect only for a side condition s; it requires there to be one more token than is necessary for any $t \in s^\bullet \cap {}^\bullet s$ to occur.

The absence of frozen tokens means that any system run will eventually either stop or involve all parts of the system. This is quite a general property since there may be a variety of reasons for a token to be frozen, ranging from its complete immobility to the fact that some infinite behaviour of the net is 'unfair'.

For finite nets, the absence of frozen tokens is stronger than safeness because of the following:

Theorem 4.3.2 [Absence of frozen token implies safeness]
Let Σ be finite and without frozen tokens. Then Σ is safe.

Proof: If Σ is unsafe then, by Theorem 3.2.6, there are transition sequences τ_0 and τ such that

$$M_0 \xrightarrow{\tau_0} M \xrightarrow{\tau} \tilde{M}$$

with $\tilde{M} > M$.

Let $|\tau| = m \in \mathbf{N}^+$ and let $s \in S$ such that $M(s) < \tilde{M}(s)$. Then starting with M, τ can be repeated $m + 2$ times to reach a marking M':

$$M_0 \xrightarrow{\tau_0} M \xrightarrow{\tau^{m+2}} M' \xrightarrow{\tau} \xrightarrow{\tau} \ldots$$

and from M' onwards there is a frozen token on s (because τ can remove at most m tokens from s). □ 4.3.2

The main objective of this section is to show that the absence of frozen tokens corresponds to the gap-freeness part of the characterisation of D-continuity. In order to cover the jump-freeness part, we first define a system equivalent of the non-single degree property:

Definition 4.3.3 [Non-single degree property for systems]
A system net $\Sigma = (S, T, F, M_0)$ is of non-single degree *iff*

$$\forall t \in T \colon (\exists M \in [\, M_0\rangle \colon M \text{ enables } t) \Rightarrow |{}^\bullet t| \neq 1 \neq |t^\bullet|.$$

$\square$ 4.3.3

The more complicated form of Definition 4.3.3 (as compared to the property given in Theorem 2.5.6) just avoids transitions that can never occur.

We have:

Theorem 4.3.4 [System equivalent of the non-single degree property]
$\Sigma = (S, T, F, M_0)$ *has the non-single degree property iff all processes of* Σ *are of non-single degree.*

Proof: We use Theorem 2.5.6 which characterises the non-single degree property for occurrence posets, and hence for occurrence nets.

($\Rightarrow$)   Consider any process $\pi = (B, E, F', p)$ of Σ which is not of non-single degree; say $|{}^\bullet e| > 1$ for some $e \in E$. Then $t = p(e) \in T$ can be enabled (i.e., $\exists M \in [\, M_0\rangle \colon M$ enables t) by Theorem 3.5.3. Further, $|{}^\bullet t| > 1$ by Property 3.3.3(iv). The case $|e^\bullet| > 1$ can be settled similarly.

($\Leftarrow$)   Consider any $t \in T$ which can be enabled, i.e., $\exists \tau$ such that τt is a transition sequence, and which satisfies $|{}^\bullet t| > 1$. In any process (B, E, F', p) corresponding to τt via Construction 3.4.1, there is an $e \in E$ with $p(e) = t$ and $|{}^\bullet e| > 1$. The case that $|t^\bullet| > 1$ is similar. $\square$ 4.3.4

We will now prove the main result of this section in three steps. Recall first that the definition of cut-boundedness requires $A \subseteq \downarrow M(A)$ (cut-boundedness from above) and $\overline{A} \subseteq \uparrow M(A)$ (cut-boundedness from below) for $A \in D$ (see Definition 2.4.13). As a first step, we will show that the latter requirement presents no problems; it is always satisfied for processes. Then we will prove a theorem which gives the process equivalent of the absence of frozen tokens. Finally, we will prove that the absence of frozen tokens and the property of non-single degree are together the system equivalent of D-continuity.

Theorem 4.3.5 [Cut-boundedness from below is always satisfied]
Let $\pi = (N, p) = (B, E, F', p)$ *be a process of a system net* $\Sigma = (S, T, F, M_0)$. *Then* $\forall A \in D(N) \colon \overline{A} \subseteq \uparrow M(A)$.

Proof: Let $A \in D(N)$; we wish to prove $\overline{A} \subseteq \uparrow M(A)$. Suppose $x \in \overline{A}$. Since $Min(N)$ is a cut, $\exists y \in Min(N): y \preceq x$. If $y \in \overline{A}$ then $y \in Min(\overline{A})$ and the result is proved. If $y \in A$ then there is an F'-chain $x_0 F' x_1 F' \dots F' x_m$ leading from $y = x_0$ to $x = x_m$, and furthermore, $\exists i: x_i \prec\!\!\cdot\; x_{i+1}$ and $x_i \in A$, $x_{i+1} \in \overline{A}$. Since $(B \cap \{x_i, x_{i+1}\}) \subseteq M(A)$ and $x_i \prec\!\!\cdot\; x_{i+1} \preceq x$, the result follows.   □ 4.3.5

Our next aim is to prove a theorem which relates the absence of frozen tokens to a property of event-infinite processes. The idea is that a token is frozen if the B-element representing it has no forward connection to the rest of the (infinite) process (as the condition labelled b in Figure 4.2). Before being able to prove this theorem, however, we will need a slightly technical theorem (and a corollary thereof) which concerns only occurrence nets and can be proved by means which have been provided in Chapter 2.

To motivate it, let us say that a B-cut c of an occurrence net N enables an event e iff $^\bullet e \subseteq c$ (it follows that $e^\bullet \cap c = \emptyset$). Suppose that N is degree-finite, discrete with respect to c and that e is some event above c, i.e., $\exists b \in c: b \prec e$. The next theorem shows that starting with c, e can be reached by finitely many steps, i.e., there is a B-cut c' such that c' enables e and between c and c' there are only finitely many events.

Theorem 4.3.6 [Events can be reached by finite steps]
Suppose that $N = (B, E, F)$ is degree-finite, discrete with respect to a B-cut c of N and that $e \in E$ is such that $\exists b \in c: b \prec e$.
Then there is a B-cut c' such that:

(i) $c \sqsubseteq c'$ (see Exercise 2 in Chapter 2);

(ii) $^\bullet e \subseteq c'$;

(iii) $|E \cap [c, c']| \in \mathbf{N}$.

Proof: By Theorem 2.2.12, we have $|[c, e]| \in \mathbf{N}$. Hence $Min(E \cap [c, e])$ is not empty and we may choose some $e_1 \in Min(E \cap [c, e])$. We claim that $^\bullet e_1 \subseteq c$. To see this, we first deduce (from $e_1 \in [c, e]$) that $b_1 \prec e_1$ for some $b_1 \in c$; we even have $b_1 \in {}^\bullet e_1$ because of the minimality property of e_1. Now suppose $b \in {}^\bullet e_1$; we cannot have $b' \prec b$ for any $b' \in c$ because of the minimality of e_1, nor can it be true that $b \prec b'$ for any $b' \in c$ because this would imply $b_1 \prec e \prec b'$, contradicting the fact that c is a co-set; the only remaining possibility is $b \in c$, since c is a cut.

Having proved that $^\bullet e_1 \subseteq c$, we are done if $e_1 = e$. Otherwise, if $e_1 \neq e$, we may define a new B-cut c_1 by the equation

$$c_1 \;=\; (c \setminus {}^\bullet e_1) \cup e_1^\bullet;$$

it is routine to check that c_1 is again a B-cut, that N is still discrete with respect to c_1, that e is above c_1 (because $e \neq e_1$) and that

$$|E \cap [c_1, e]| \;=\; |E \cap [c, e]| - 1.$$

The construction can be repeated until eventually e is in $Min(E \cap [c_i, e])$ for one of the B-cuts c_i so constructed; it is guaranteed to end after at most $|E \cap [c, e]| - 1$ steps yielding some $c' = c_i$ with ${}^\bullet e \subseteq c'$. Properties (i) and (iii) of the theorem follow immediately from the construction. $\square$ 4.3.6

The construction given in Theorem 4.3.6 works in such a way that the part of c which is not in $[c, e]$ remains untouched. In particular, if b_0 is a condition in c which is co to e then b_0 is also in c'. We formulate this as a corollary:

Corollary 4.3.7
With the same notation as in Theorem 4.3.6, suppose $b_0 \in c$ such that b_0 co e. Then c' can be chosen such that, in addition to the properties stated in Theorem 4.3.6, $b_0 \in c'$ holds true.

Proof: This is a property of the construction used to prove Theorem 4.3.6.
     $\square$ 4.3.7

Now we are ready to prove the second step of the main result of this section:

Theorem 4.3.8 [Absence of frozen tokens]
Let $\Sigma = (S, T, F, M_0)$ be a finite system.
Then Σ has no frozen tokens iff all event-infinite processes $\pi = (B, E, F', p)$ of Σ satisfy the property

$$\boxed{*}\quad \forall x, y \in B \cup E: \uparrow x \cap \uparrow y \neq \emptyset.$$

Proof:

($\Rightarrow$)   Suppose $\pi = (N, p) = (B, E, F', p)$ is an event-infinite process of Σ which does not satisfy property $\boxed{*}$. Using the fact that $Min(N)$ is a finite cut, degree-finiteness and the infinity of N, we may construct an infinite li-set

$$x_0 \prec x_1 \prec x_2 \prec \ \ldots$$

with $x_0 \in Min(N)$ (exactly as in the proof of Theorem 2.2.10). Since N fails to have property $\boxed{*}$, there is some $x \in B \cup E$ and some index $j_0 \geq 0$ such that x co x_j for all $j \geq j_0$ (otherwise $\uparrow x$ and $\uparrow y$ would always meet on some point of the li-set). If $x \in E$ then $b \in x^\bullet$ has the same property. Hence, without loss of generality we may assume $x \in B$.

Define $s = p(x)$. Further, define $b_1 = Min_{j \geq 0}\{x_j \in B \mid x_j \text{ co } x\}$ (see Figure 4.3).

Starting with b_1 we construct an 'infinite execution which is co to x'. Since $\{x, b_1\}$ is a finite B-co-set, by Theorem 3.5.8 we find a reachable B-cut c_0 which contains x and b_1, i.e., $\{x, b_1\} \subseteq c_0$. Since $A = \{x_0, x_1, \ldots\}$ is an infinite li-set and $b_1 \in A$, there are infinitely many events above b_1 in A. By resuming the

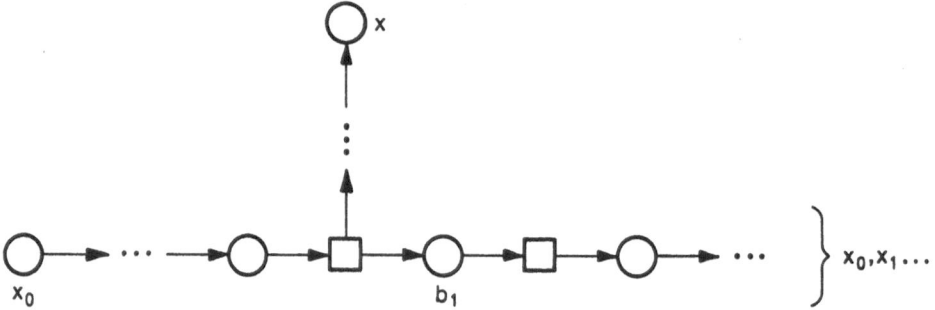

Fig. 4.3. Illustration of the proof of ($\Rightarrow$)

construction given in Theorem 4.3.6 repeatedly, we find events e_i and B-cuts c_i such that $c_i = (c_{i-1} \setminus {}^\bullet e_i) \cup e_i^\bullet$ for $i \geq 1$ (not necessarily are all e_i in A, but they satisfy, as is easy to see, x co e_i). By Corollary 4.3.7, the c_i satisfy $x \in c_i$ for all $i \geq 0$. Let M_0' correspond to c_0 and M_i correspond to c_i ($i \geq 1$) according to Definition 3.4.8, and put $t_i = p(e_i)$ for $i \geq 1$. Then $M_0' \in [\, M_0\rangle$ and $M_0' t_1 M_1 t_2 M_2 \ldots$ is an occurrence sequence which satisfies the desired properties with a frozen token on $s = p(x)$.

($\Leftarrow$)   Suppose that Σ has a frozen token. That is, there is an infinite occurrence sequence

$$\sigma = M_0 t_0 \ldots M_1 t_1 M_2 t_2 \ldots$$

and a place $s \in S$ such that for all $i \geq 1$: $M_i(s) \geq 1$ if $s \notin {}^\bullet t_i$ and $M_i(s) \geq 2$ if $s \in {}^\bullet t_i$.

Using Construction 3.4.1 we define an event-infinite process associated to σ as follows. First note that $\sigma = \sigma_1 \sigma_2$ where σ_1 leads from M_0 to M_1 and σ_2 starts with M_1. For σ_1 we take any of the processes that may result by an application of Definition 3.4.1; say this is (N_1, p_1). We know that $Max(N_1)$ is a B-cut which corresponds to M_1. We now construct a process (N, p) by induction on $i \geq 1$. For $i = 1$ let (N_1, p_1) be given as above. Since $M_1(s) \geq 1$, we can fix some $b_0 \in Max(N_1)$ such that $p_1(b_0) = s$ and $b_0^\bullet = \emptyset$. Suppose (N_i, p_i) to be given; we construct (N_{i+1}, p_{i+1}) as in Definition 3.4.1, except for the following. If $s \in {}^\bullet t_i$ then $M_i(s) \geq 2$ and there exists $b_0' \neq b_0$, $b_0' \in Max(N_i)$ and $p_i(b_0') = s$; we use b_0' (rather than b_0) when adding a new event e_i to N_i, so that in N_{i+1} we still have $b_0^\bullet = \emptyset$. If, on the other hand, $s \notin {}^\bullet t_i$ then b_0 need not be used in this construction, whence again $b_0^\bullet = \emptyset$ in N_{i+1}. Hence $b_0^\bullet = \emptyset$ in N and therefore, $\uparrow b_0 \cap \uparrow e = \emptyset$ for any event e of N not in N_1. Hence (N, p) is an event-infinite process which does not have property $\boxed{*}$. $\square$ 4.3.8

Theorem 4.3.9 [The system equivalent of D-continuity]

Let $\Sigma = (S, T, F, M_0)$ be a finite system.
Then Σ is of non-single degree and has no frozen tokens iff for all processes
$\pi = (N, p) = (B, E, F', p)$ *of Σ, N is D-continuous.*

Proof: We shall use Theorem 2.5.8 which states that an occurrence net is
D-continuous iff it is K-dense, cut-bounded and of non-single degree.

($\Rightarrow$)   Let $\pi = (N, p)$ be a process of a system Σ which is of non-single degree
and has no frozen tokens.

From Theorem 4.3.4($\Rightarrow$) it follows that N is of non-single degree.

From Theorem 4.3.2 it follows that Σ is safe; hence, by Theorem 4.2.1($\Rightarrow$),
N is K-dense.

From Theorem 4.3.5 it follows that N is cut-bounded from below.

If N is event-finite then $Max(N)$ is a B-cut, and cut-boundedness from
above follows by an argument dual to the one employed in Theorem 4.3.5.

Assume that N is event-infinite. By Theorem 4.3.8, it follows that

$$\boxed{*} \quad \forall x, y \in B \cup E : \uparrow x \cap \uparrow y \neq \emptyset.$$

Using this fact, we shall prove that N is cut-bounded from above, i.e.,

$$\forall A \in D : A \subseteq \downarrow M(A).$$

Let $A \in D$ and $x \in A$. We have to prove that there exists an element $y \in M(A)$
such that $x \preceq y$. Since $\overline{A} \neq \emptyset$, we can choose $z \in \overline{A}$. By Property $\boxed{*}$, it follows
that $\exists u \in \overline{A} : x \prec u \wedge z \preceq u$. By combinatorialness, there exists a finite sequence
$x_1, x_2, \ldots, x_m \in B \cup E$ such that:

$$x = x_1 \prec x_2 \prec \ldots \prec x_m = u.$$

$x \in A \wedge u \in \overline{A}$ implies that $\exists i \in \mathbf{N}^+$, $1 \leq i < n$, such that $x_i \in A \wedge x_{i+1} \in \overline{A}$.

Claim: Either $x_i \in Max(A)$ or $x_{i+1} \in Min(\overline{A})$ or both.

Proof: (Indirectly.) Assume that $x_i \notin Max(A) \wedge x_{i+1} \notin Min(\overline{A})$.
Then $\exists x' \in A : x' \succ x_i$ and $\exists x'' \in \overline{A} : x'' \prec x_{i+1}$.

It follows easily that $x_i \ co \ x'' \ co \ x' \ co \ x_{i+1}$. By N-density, there exists
$w \in X$ with $x_i \prec w \prec x_{i+1}$, a contradiction since $x_i \prec x_{i+1}$. $\square$ *Claim*

By the claim and since $x \preceq x_i \prec x_{i+1}$, it follows that $x \in \downarrow M(A)$, i.e., N is
cut-bounded from above.

($\Leftarrow$)   Let all processes (N, p) of Σ be K-dense, cut-bounded and of non-single
degree.

The non-single degree property of Σ follows from Theorem 4.3.4($\Leftarrow$).

To complete the proof, we will show that from K-density, cut-boundedness,
degree-finiteness and the fact that $Min(N)$ is a finite cut it follows that all

infinite processes of Σ satisfy Property $\boxed{*}$; the fact that Σ has no frozen tokens then follows with Theorem 4.3.8($\Leftarrow$).

As in the proof of Theorem 4.3.8($\Rightarrow$), from our assumptions we may find an infinite li-set

$$x_0 \prec\!\!\!\!\cdot\ \ x_1 \ \prec\!\!\!\!\cdot\ \ x_2 \ \prec\!\!\!\!\cdot\ \ \ldots$$

with $x_0 \in Min(N)$.

We claim that $\forall x \in B \cup E\ \exists j\colon x \preceq x_j$.
From this, Property $\boxed{*}$ of Theorem 4.3.8 follows because if $x \preceq x_j$ and $y \preceq x_k$ then $x_{\max(j,k)} \in {\uparrow}x \cap {\uparrow}y$.

To prove the claim, suppose that $\exists x \in B \cup E\ \forall j\colon \neg(x \preceq x_j)$; we will arrive at a contradiction. Then there exists an index j_0 such that $x\ co\ x_j$ for $j \geq j_0$ (see Figure 4.4).

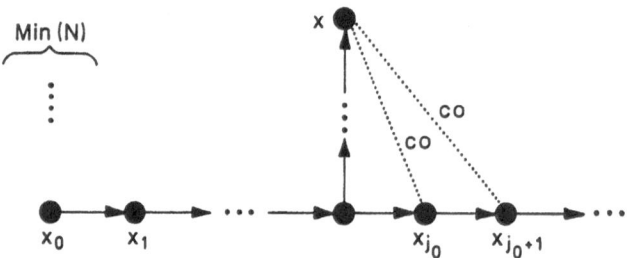

Fig. 4.4. Illustration of the proof of ($\Leftarrow$)

Now define $A = {\downarrow}\{x_0, x_1, x_2, \ldots\}$; $x \in \overline{A}$ by the choice of x. We also have $A \in D(N)$. Now suppose that $\forall x_i\ \exists y \in \overline{A}\colon x_i \prec y$. Then a line containing $\{x_0, x_1, x_2, \ldots\}$ cannot meet a cut containing $Min(\overline{A})$, contradicting K-density.

Hence $\exists x_i\ \forall y \in \overline{A}\colon x_i\ co\ y$. Fix such an x_i. By the definition of x_i, $x_i \notin {\downarrow}Min(\overline{A})$. But also $x_i \notin {\downarrow}Max(A)$ because $Max(A) = \emptyset$ by the definition of A. Hence $x_i \notin {\downarrow}M(A)$, contradicting cut-boundedness from above.

This yields the desired contradiction and hence proves the original claim, and with it the theorem. $\square$ 4.3.9

Recall from Section 2.4 that D-continuity is equivalent to jump-freeness and gap-freeness. From the characterisation theorems 2.4.10 and 2.4.12 and from Theorem 4.3.9, we may deduce that the non-single degree property is the system equivalent of jump-freeness and the absence of frozen tokens is the system equivalent of gap-freeness (for finite systems, of course). The absence of frozen tokens is an interesting well-behavedness property of systems with some applications, the discussion of which, however, falls outside the scope of this monograph.

4.4 A Closing Remark on Finite 1-safe Nets

In this section, we restate some of the main results of this book in basic terms for one of the most frequently used classes of Petri nets: finite 1-safe systems. In the following, let $\Sigma = (S, T, F, M_0)$ be finite ($|S \cup T| \in \mathbf{N}$) and 1-safe ($\forall M \in [M_0\rangle \; \forall s \in S : M(s) \leq 1$).

- Every occurrence sequence of Σ has exactly one associated process (Theorem 3.4.9).

- Every process of Σ has at least one associated occurrence sequence (Theorem 3.5.3).

- Every process of Σ is combinatorial (since it is an occurrence net), N-dense (Theorem 2.5.4), countable (Theorem 3.3.4) and of finite degree (Remark 3.5.2).

- Every process of Σ is K-dense (Theorem 4.2.1) and hence, by the results of Sections 2.2 and 2.3, weakly discrete (Theorem 2.3.6(i)$\Rightarrow$(ii)), interval-finite (Theorem 2.2.10(iii)), boundedly discrete (Theorem 2.2.10(i)) and injectively observable (Theorem 2.2.16($\Leftarrow$)).

- Every B-cut of every process of Σ is finite and reachable (this follows with K-density from Theorem 2.3.12 and Theorem 3.5.8).

- Every process of Σ is discrete with respect to every B-cut c and observable with respect to every B-cut c (not just the initial cut). (See Exercise 6.) Thus, the processes of Σ satisfy fully all properties investigated in Sections 2.2 and 2.3.

- If c and c' are two B-cuts of a process of Σ such that $c \sqsubseteq c'$, then there is a finite sequence of B-cuts $c_0, \ldots, c_m$ and a finite sequence of events $e_1, \ldots, e_m$ such that $c_0 = c$, $c_m = c'$ and $\forall i, 1 \leq i \leq m : c_i = (c_{i-1} \backslash {}^\bullet e_i) \cup e_i^\bullet$, i.e., c' is reachable from c by a finite number of single events. (This follows from the fact that both c and c' are reachable, exactly as in the proof of Theorem 3.5.7).

- Every process of Σ is cut-bounded from below (Theorem 4.3.5).

- A process of Σ is gap-free iff it is cut-bounded from above (Theorems 2.4.12 and 2.5.7).

Exercises.

1. * Let Σ be a finite system net.
   Prove that for all processes $\pi = (N, p)$ of Σ: if c is a reachable B-cut of π and l is a line of π then $c \cap l \neq \emptyset$.
   (Hint: This is a restricted kind of K-density, showing that non-K-dense processes always involve non-reachable B-cuts; use arguments as in Chapter 2.)

2. Does the statement in Exercise 1 hold for infinite Σ / for Σ of finite synchronisation / for g-reachable B-cuts (see Exercise 12 of Chapter 3)?

3. Do the following system nets have any frozen tokens?

   (i)   Figure 3.1(i).

   (ii)   Figure 3.6(i).

   (iii) Figure 3.8(i).

4. Show that Theorem 4.3.2 is wrong for infinite Σ.

5. Does Theorem 4.3.5 hold for infinite Σ?

6. Let Σ be a finite 1-safe system, π a process of Σ and c an arbitrary B-cut of π. Prove that π is discrete with respect to c.

Bibliographical Notes

References to Chapter 1. Introduction

The work that initiated Petri net theory is C.A.Petri's dissertation [39].

The nontransitivity of concurrency has been stressed frequently by C.A.Petri. It has also been tacitly and widely acknowledged by many computer scientists, starting, perhaps, with the discussion in E.W.Dijkstra's article [14] and expressed, perhaps, most prominently by L.Lamport in [31].

Petri's seminal papers that contain the properties we investigate are [41], [43] and [44].

A.W.Holt's report [27] also contains ideas and material we have used in this monograph.

The final report of the project BEGRUND at the Gesellschaft für Mathematik und Datenverarbeitung by E.Best, C.Fernández and H.Plünnecke [6] has been a direct precursor of this book.

There are several important, alternative approaches to the study of the nonsequential behaviour of a system which are related to partially ordered sets. We mention the theory of traces initiated by A.Mazurkiewicz [34], the calculus of vector firing sequences devised by M.W.Shields [53], the theory of event structures first studied by M.Nielsen, G.Plotkin and G.Winskel [35] and further developed by G.Winskel (see the survey paper [60]) and the theory of partial words and semiwords introduced by J.Grabowski [24] and P.Starke [56], respectively. All of these approaches are event-based, contrasting with the approach of J.Winkowski [57] which is state-based.

References to Chapter 2. Partial Orders

Section 2.1. Introduction and Basic Definitions

Chains and antichains are basic concepts in several mathematical theories, for instance, in the theory of partial orders, in lattice theory and in the theory of graphs. Some fundamental textbooks in this area are [1,8,30,37].

The terminology to let 'lines' and 'cuts' denote maximal chains and maximal antichains, respectively, is due to C.A.Petri [41].

Section 2.2. Combinatorialness and Discreteness

Density and combinatorialness are again basic concepts in Mathematics and, in particular, in the theory of partial orders. In [43], C.A.Petri mentions combinatorialness as a basic property of occurrence nets.

Weak discreteness is closely related with the idea of putting some restriction on the length of maximal chains between two points and is thus related – without being the same – to old thoughts in Mathematics such as the Jordan-Dedekind chain condition. In computer science, the concept of weak discreteness first appears (as far as we know) in a paper by C.Hewitt and H.Baker [26]. In [2], E.Best proves that K-dense occurrence nets are weakly discrete.

Bounded discreteness – also found in general Mathematics, albeit with different names (see, for instance, [1]) – has first been mentioned in conversations by G.D.Plotkin and G.Winskel and is defined in [58,59].

The notion of observability – which is related to the mathematical concept of 'graded posets' [8] – and Theorem 2.2.15 which links observability to bounded discreteness are due to G.Winskel [58].

Interval-finiteness was encountered as a useful proof concept independently by many researchers. Its connection 2.2.16 to injective observability was discovered by E.Best [4] and E.Smith [54]. Degree-finiteness is another basic concept of graph theory. Like countability, degree-finiteness is a typical 'precondition'; that is to say, it allows various results to be proved but is not itself the consequence of other properties.

Discreteness with respect to a cut has been defined – albeit less prominently than here – by G.Winskel in [58] and by C.Fernández, M.Nielsen and P.S.Thiagarajan in [17]. We have decided to stress the importance of this concept because it turns out to be a property needed in the definition of a process.

The correspondence between discreteness on the one hand and observability on the other hand can be augmented by a correspondence with a property known as generability. For the details, the reader is referred to the paper [17] by C.Fernández, M.Nielsen and P.S.Thiagarajan.

Section 2.3. N-density and K-density

N-density has been introduced by C.A.Petri [41] in connection with K-density. K-density has also been defined in [41]. The concept itself is older: R.Carnap has derived K-density in his axiom system of Physics [10, T35 in Appendix G]. A.W.Holt has proved that finite occurrence nets are K-dense [27].

Independently of Petri's work, the concepts of N-density and K-density (with different names) have been used in some mathematical papers; for instance, the reader may be referred to a paper by P.A.Grillet [25].

The concept of embeddability is used by E.Best [2] to prove a characterisation of K-density. Theorem 2.3.6 and its proof are cleaned versions and mild
generalisations of the results of [2].

H.Plünnecke has provided a considerable generalisation of Theorem 2.3.6
[47,48]; he has shown that combinatorialness is not, in fact, needed to prove
Proposition 2.3.9 and that a poset is K-dense provided that it is N-dense and
has only finite cuts (that is, a strengthening of Corollary 2.3.11).

The curious reader may consult the report by E.Best, C.Fernández and
H.Plünnecke [6] which gives an overview of the general results without stating
their proofs in detail.

A different view of K-density – in terms of the relation $\sqsubseteq$ defined in Exercise 2 of Chapter 2 – is described by C.Fernández and P.S.Thiagarajan [19].

Section 2.4. D-continuity

The definition of D-continuity is taken from C.A.Petri's papers [43,44]. This
concept has been inspired by R.Carnap's axiom system described in [10, Appendix G] and, of course, by R.Dedekind's seminal paper [12].

C.Fernández and P.S.Thiagarajan discuss the implications of this definition
in [18] where necessary conditions for D-continuity are derived and D-continuity
is characterised for a class of occurrence nets. These results are extended
by E.Best and A.Merceron to yield a characterisation of D-continuous posets;
see [7] for an overview. The non-single degree property and cut-boundedness
have been identified in the course of these efforts. At the same time, it was
realised that the two parts of D-continuity, i.e., jump-freeness and gap-freeness,
can be characterised separately.

Variants of the original definition of D-continuity have been discussed by
C.A.Petri and E.Smith [46].

Section 2.5. Occurrence Posets

Occurrence nets are central to the description of concurrent behaviour. Occurrence posets are simply the class of posets that correspond to occurrence nets.
Their characterisation 2.5.3 allows simpler proofs of some of their properties
and shows that combinatorialness is an independent and essential aspect of
their definition.

Theorem 2.5.4 belongs to the 'net folklore', while the remaining theorems
of Section 2.5 are simple consequences of the general results of the preceding
sections.

Another reasonable way of deriving a poset from an occurrence net (B, E, F)
is to omit all elements of B, that is, to consider the poset $(E, F^+|_{E \times E})$. The
latter objects are similar to 'occurrence graphs' introduced by A.W.Holt [27]
and to 'elementary event structures' [35].

Another poset that may be derived from an occurrence net $N = (B, E, F)$ is $(B, F^+|_{B \times B})$, that is, the sub-poset spanned by the conditions of N. For a detailed study of similar kinds of objects, the reader is referred to [57].

Our usage of the term 'occurrence net' dates back to C.A.Petri who, from [42] onwards, has used this term as a translation of the German word Kausalnetz; in [40] he has used the term 'occurrence structure' while in the English version of [41] we find 'causal net'. Independently of this, [35] have defined 'occurrence nets' to be a more general (and also very important) class of nets; the reader should be careful not to confuse the two definitions.

Exercises to Chapter 2

The relation defined in Exercise 2 is studied by C.Fernández and P.S.Thiagarajan [19].

E.Smith [54] proves the statement of Exercise 9.

An (uncountable) poset which is boundedly discrete but not observable has been constructed by G.Winskel [58].

The solutions to Exercises 25 and 26 can be found in [7].

References to Chapter 3. Petri Nets

Section 3.1. Nets and Markings

The Petri net model we are considering is simple, basic and well-established.

Possible generalisations include allowing arbitrary (finite) arc weights (causing a transition to remove as many tokens from a place as defined by the weight), allowing individual tokens, and allowing limits on the number of tokens on a place. For these (and other) generalisations, the reader is referred to the standard textbooks by G.W.Brams [9], J.L.Peterson [38], W.Reisig [49] and P.H.Starke [55], to the general literature, and to the terminology compiled by E.Best and C.Fernández [5].

Our Petri net model is basic in the sense that it may serve as the target of translations from many other models of concurrent systems (including concurrent programs, as shown by E.Best [3] and by P.Degano, R.Gorrieri and S.Marchetti [13]). The model is also established in the sense that it has been the subject of widespread study.

We do not limit our considerations to finite Petri nets because infinite Petri nets may arise in a variety of circumstances, as discussed by E.Best and R.Devillers [4].

Section 3.2. Transition Rule and Occurrence Sequences

The transition rule is fundamental to any semantics of Petri nets. It consists of two parts: the enabling condition and the calculation of a successor marking. In the literature, variants of both parts have been discussed, either to capture capacity constraints, or to describe probabilities, or for other reasons. However, for the class of nets considered here, the rule we have given is generally accepted.

Occurrence sequences and transition sequences are the standard semantics of marked Petri nets defined in the textbooks [9,38,49] and in the contribution by H.J.Genrich, K.Lautenbach and P.S.Thiagarajan [21] to the first Advanced Course on Petri Nets. They describe (finitely or infinitely often) repeated applications of the transition rule, that is, the 'arbitrary interleaving' semantics of Petri nets.

The notion of concurrent enabling has been defined and discussed, amongst others, by H.J.Genrich and E.Stankiewicz-Wiechno [22], U.Goltz and W.Reisig [23,50] and by E.Best and R.Devillers [4].

Safeness (often also called boundedness) is one of the oldest established properties of Petri nets; its origins date back to work by H.J.Genrich [20] and F.Commoner and A.W.Holt [28].

Theorem 3.2.6 is a piece of standard theory of Petri nets; the arguments are similar, for instance, to those employed in Theorem 4.1 of J.L.Peterson's textbook [38].

Section 3.3. Occurrence Nets and Processes

The process properties (Definition 3.3.3) originate from C.A.Petri's seminal paper [41] where four properties have been described informally. The first formalisation of these properties was done by H.J.Genrich, K.Lautenbach and P.S.Thiagarajan [21].

Their definition has been cleaned and generalised (to not necessarily safe nets and to arbitrary arc weights) by U.Goltz and W.Reisig [23]. The latter have also noticed (with the help of Example 3.4) that some form of discreteness should be included in the list of process properties.

E.Best and R.Devillers [4] have argued, by proving consistency results with inductive definitions, that the set of properties as given in Definition 3.3.3 (including discreteness with respect to the initial cut) is exactly what is needed for the class of nets under consideration here.

Process properties for other classes of nets have been given by E.Best, C.Fernández and L.Castellano [5,11] (for condition/event systems) and by G.Rozenberg [51] (for elementary net systems). If the conditions of a process are omitted then one gets a labelled partial order on events. It is possible (and done by P.H.Starke [56]) to give a direct definition of labelled partial orders of events as the concurrent behaviour of marked nets. The relationship between these objects and the processes as defined in the present section is investigated, for the class of nets we consider, by A.Kiehn [29].

Section 3.4. Inductive Definition of Processes

Construction 3.4.1 can be found in this form by E.Best and A.Merceron [7] where it is used for proof purposes. It is also part of a proof in the paper [23] by U.Goltz and W.Reisig. E.Best and R.Devillers have used various generalisations of this construction to justify the process properties [4]. A different inductive definition of processes can be found in the survey paper by M.Nielsen and P.S.Thiagarajan [36].

Proposition 3.4.7 is from [23].

Theorem 3.4.9 has an interesting weak converse: the processes of a safe system are uniquely characterised by their linearisations; for this result, the reader is referred to the work by A.Mazurkiewicz [34] (for finite processes) and by E.Best and R.Devillers [4] (for infinite processes).

Section 3.5. Systems of Finite Synchronisation

The terminology has been introduced by E.Best and R.Devillers; see [4] where more facts about the systems of finite synchronisation are proved.

Bounded nondeterminism has been discussed by E.W.Dijkstra [15] and others.

Theorem 3.5.3 is in [4].

In [4,6] it is shown that concurrent programs can be translated into Petri nets which are of finite synchronisation.

Theorem 3.5.7 is from [4], while Theorem 3.5.8 is in [7].

Exercises to Chapter 3

Exercise 10 has been discussed in [4].

References to Chapter 4. Connections Between Systems and Processes

Section 4.2. K-density and Safeness

Theorem 4.2.1 is due to U.Goltz and W.Reisig [23].

Section 4.3. D-continuity and Frozen Tokens

The results proved in this chapter are due to E.Best and A.Merceron [7]. The significance of frozen tokens is discussed further by W.M.Lu and A.Merceron [32,33].

Section 4.4. A Closing Remark on Finite 1-safe Nets

The facts stated in this section for finite 1-safe net systems apply in particular for two Petri net classes which are widely recognised as basic in the literature. Finite contact-free condition-event systems [21] and finite contact-free elementary net systems [52] are both special classes of finite 1-safe net systems, and hence, all the results stated in Section 4.4 are valid for these systems.

Exercises to Chapter 4

Exercise 1 has been discussed by G.Winskel [58].

Bibliography

[1] BERGE, C.: *The Theory of Graphs and its Applications*. First published by Dunod, Paris (1958); English translation: Methuen and Co. (1962).

[2] BEST, E.: *A Theorem on the Characteristics of Non-sequential Processes*. University of Newcastle upon Tyne, Computing Laboratory, Tech. Report 116 (Nov., 1977)
also: Annales societas mathematicae polonae Ser. IV Fundamenta Informaticae III.1, pp. 77–94 (1980)

[3] BEST, E.: *COSY: Its Relation to Nets and to CSP*. Petri Nets: Applications and Relationships to Other Models of Concurrency, Advances in Petri Nets 1986, Part II, Proceedings of an Advanced Course, Bad Honnef, September 1986, Brauer, W.; Reisig, W.; Rozenberg, G. (eds.), Lecture Notes in Computer Science Vol.255, — Springer-Verlag, pp.416-440 (1987)

[4] BEST, E.; DEVILLERS, R.: *Concurrent Behaviour: Sequences, Processes and Programming Languages*. Gesellschaft für Mathematik und Datenverarbeitung Bonn, GMD-Studien Nr. 99 (May, 1985) The net theoretical part has appeared as: *Sequential and Concurrent Behaviour in Petri Net Theory*. Theoretical Computer Science, Vol.55, No.1 (1988)

[5] BEST, E.; FERNÁNDEZ, C.: *Notations and Terminology on Petri Net Theory*. Gesellschaft für Mathematik und Datenverarbeitung, Bonn, Arbeitspapiere der GMD Nr. 195 (Jan., 1986)

[6] BEST, E.; FERNÁNDEZ, C.; PLÜNNECKE, H.: *Concurrent Systems and Processes. Final Report of the Foundational Part of the Project BEGRUND*. Gesellschaft für Mathematik und Datenverarbeitung , Bonn, GMD-Studien Nr. 104 (March, 1985)

[7] BEST, E.; MERCERON, A.: *D-Continuity and Petri's Axioms of Concurrency for Nonsequential Process Models*. Fundamenta Informaticae X, pp.161-212 (1987)

[8] BIRKHOFF, G.: *Lattice Theory*. Published by the American Mathematical Society. First edition (1940)

[9] BRAMS, G.W. nom collectif de Ch. André, G. Berthelot, C. Girault, G. Memmi, G. Roucairol, J. Sifakis, R. Valette, G. Vidal-Naquet: *Réseaux de Petri: Théorie et Pratique* Tome. 1 – Théorie et Analyse. Tome 2 - Modélisation et Applications. Editions Masson (Sep., 1982)

[10] CARNAP, R.: *Symbolische Logik.* 2. Auflage, Springer Verlag, Wien (1960)

[11] CASTELLANO, L.A.: *Beta-Processes of C/E Systems.* Lecture Notes in Computer Science Vol. 222: Advances in Petri Nets 1985 / Rozenberg, G. (ed.) — Springer Verlag, pp. 83–100 (1986)

[12] DEDEKIND, R.: *Was sind und was sollen die Zahlen? — Stetigkeit und irrationale Zahlen.* Braunschweig (1887)

[13] DEGANO, P., GORRIERI, R. AND MARCHETTI, S.: *An Exercise in Concurrency: A CSP Process as a Condition/Event System.* Proceedings of the Eighth European Workshop on Application and Theory of Petri Nets, Zaragoza, Spain, pp.31-50 (June, 1987).

[14] DIJKSTRA, E.W.D.: *Cooperating Sequential Processes.* Programming Languages: Genuys,F. (ed.), Academic Press, New York, pp.43-112 (1968).

[15] DIJKSTRA, E.W.D.: *A Discipline of Programming.* Prentice Hall (1976).

[16] DREES, S.; GOMM, D.; PLÜNNECKE, H.; REISIG, W.; WALTER, R.: *Bibliography of Net Theory.* Arbeitspapiere der GMD Nr. 212 (June 1986).

[17] FERNÁNDEZ, C.; NIELSEN, M.; THIÁGARAJAN, P.S.: *Notions of Realizable Non-Sequential Processes.* Fundamenta Informaticae IX, pp.421-454 (1986)

[18] FERNÁNDEZ, C.; THIAGARAJAN, P.S.: *D-Continuous Causal Nets: A Model of Non-sequential Processes.* Gesellschaft für Mathematik und Datenverarbeitung, ISF-Report 82.05 (Juli, 1982)
also: Theoretical Computer Science 28, pp. 171–196 (1984)

[19] FERNÁNDEZ, C.; THIAGARAJAN, P.S.: *A Lattice Theoretic View of K-Density.* Gesellschaft für Mathematik und Datenverarbeitung Bonn, Arbeitspapiere der GMD Nr. 76 (Dez., 1983)
also: Lecture Notes in Computer Science Vol. 188: Advances in Petri Nets 1984 / Rozenberg, G. (ed.) — Springer-Verlag, pp. 139–153 (1985)

[20] GENRICH, H.J.: *Das Zollstationenproblem.* Gesellschaft für Mathematik und Datenverarbeitung Bonn, Interner Bericht ISF/69–01–15 (1969)
also: Revised Version: GMD, ISF/71–10–13 (1971)

[21] GENRICH, H.J.; LAUTENBACH, K.; THIAGARAJAN, P.S.: *Elements of General Net Theory.* Lecture Notes in Computer Science Vol. 84: Net Theory and Applications, Proceedings of the Advanced Course on General

Net Theory of Processes and Systems, Hamburg, 1979 / Brauer, W. (ed.) — Berlin, Heidelberg, New York: Springer-Verlag, pp. 21–163 (1980)

[22] GENRICH, H.J.; STANKIEWICZ-WIECHNO, E.: *A Dictionary of some Basic Notions of Net Theory*. Lecture Notes in Computer Science Vol. 84: Net Theory and Applications, Proceedings of the Advanced Course on General Net Theory of Processes and Systems, Hamburg, 1979 / Brauer, W. (ed.) — Berlin, Heidelberg, New York: Springer-Verlag, pp. 519–535 (1980)

[23] GOLTZ, U.; REISIG, W.: *The Non-sequential Behaviour of Petri Nets*. Information and Control Vol. 57, No. 2–3, pp. 125–147 (May-June, 1983)

[24] GRABOWSKI, J.: *On Partial Languages*. Fundamenta Informaticae IV.2 (1981).

[25] GRILLET, P.A.: *Maximal chains and antichains*. Fundamenta Mathematicae, LXV (1969), pp.157-167.

[26] HEWITT, C.; BAKER, H.: *Actors and Continuous Functionals*. In: Formal Description of Programming Concepts/ Neuhold, E.J. (ed.), North Holland (1978)

[27] HOLT, A.W. et al.: *Information System Theory Project: Final Report*. Princeton, N. J.: Applied Data Research, Inc., RADC-TR-68–305, NTIS AD 676972 (Sep., 1968)

[28] HOLT, A.W.; COMMONER, F.: *Events and Conditions*. Princeton, N. J.: Applied Data Research Inc., Information System Theory Project (1970)

[29] KIEHN, A.: *On the Interrelation Between Synchronized and Non-synchronized Behaviour of Petri Nets*. Elektronische Informationsverarbeitung und Kybernetik (EIK) 1-2 (1988).

[30] KÖNIG, D.: *Theorie der endlichen und unendlichen Graphen*. Akademische Verlagsgesellschaft, Leipzig (1936). Reprinted: Chelsea Publishing Company, New York (1950)

[31] LAMPORT, L.: *Time, Clocks, and the Ordering of Events in a Distributed System*. Communications of the ACM, 21(7), pp.558-565 (1978).

[32] LU, W.M.; MERCERON, A.: *The Meaning of Frozen Tokens*. Rapport de Recherche No. 218 L.R.I., U.A. au CNRS 410 "Al Khowarizmi", Bat. 490 Univ. Paris XI, 91405 Orsay Cedex France (Mai, 1985) also: Proceedings ISCAS '85, Kyoto, Japan. — New York: IEEE, pp. 495–498 (1985)

[33] LU, W.M.; MERCERON, A.: *Frozen Tokens - A Way to Test Distributed Systems Modeled by Petri Nets.* Institute of Mathematics, Academia Sinica, Beijing, People's Republic of China, Research Memorandum, C. No. 2 (June, 1985)

[34] MAZURKIEWICZ, A.: *Concurrent Program Schemes and Their Interpretation.* Aarhus University, Computer Science Department, DAIMI PB-78 (July, 1977)

[35] NIELSEN, M.; PLOTKIN, G.; WINSKEL, G.: *Petri Nets, Event Structures and Domains.* Lecture Notes in Computer Science Vol. 70: Semantics of Concurrent Computation / Kahn, G. (ed.) — Berlin: Springer-Verlag, pp. 266–284 (1979)

[36] NIELSEN, M.; THIAGARAJAN, P.S.: *Degrees of Non-Determinism and Concurrency: A Petri Net View.* Lecture Notes in Computer Science Vol.181. Foundations of Software Technology and Theoretical Computer Science. Fourth Conference, Bangalore, India, December 1984. Proceedings. Joseph, M.; Shyamasundar, R. (eds.), Springer Verlag, pp. 89–117 (1984)

[37] ORE, Ø.: *Theory of Graphs.* American Mathematical Society Colloquium Publications Vol.38, Rhode Island (1962)

[38] PETERSON, J.L.: *Petri Net Theory and the Modeling of Systems.* Englewood Cliffs, New Jersey: Prentice Hall, Inc. (1981)

[39] PETRI, C.A.: *Kommunikation mit Automaten.* Bonn: Institut für Instrumentelle Mathematik, Schriften des IIM Nr. 2 (1962)
also: New York: Griffiss Air Force Base, Technical Report RADC-TR-65-377, Vol.1, Suppl. 1 ⟨English translation⟩ (1966)

[40] PETRI, C.A.: *Concepts of Net Theory.* Mathematical Foundations of Computer Science: Proceedings of Symposium and Summer School, High Tatras, Sep. 3–8, 1973. — Mathematical Institute of the Slovak Academy of Sciences, pp. 137–146 (1973)

[41] PETRI, C.A.: *Nicht-sequentielle Prozesse.* Universität Erlangen-Nürnberg, Arbeitsberichte des IMMD, Vol.9, Nr.8, pp. 57–82 (1976)
also: Gesellschaft für Mathematik und Datenverarbeitung Bonn, ISF-Bericht ISF-76-6, 3., revidierte und ergänzte Auflage (15.06.1977). Translation: *Non-Sequential Processes: Translated by Philip Krause and John Low.* Gesellschaft für Mathematik und Datenverarbeitung Bonn, Interner Bericht ISF-77-5 (1977)

[42] PETRI, C.A.: *Concurrency as a Basis of Systems Thinking.* Gesellschaft für Mathematik und Datenverarbeitung Bonn, Interner Bericht ISF-78-06 (Sep., 1978)

also: Proceedings from 5th Scandinavian Logic Symposium, Jan., 1979, Aalborg / Jensen, F.V.; Mayoh, B.H.; Moller, K.K. (eds.) — Aalborg: Universitetsforlag, pp. 143–162 (1979)

[43] PETRI, C.A.: *Concurrency.* Lecture Notes in Computer Science Vol. 84: Net Theory and Applications, Proceedings of the Advanced Course on General Net Theory of Processes and Systems, Hamburg, 1979 / Brauer, W. (ed.) — Berlin, Heidelberg, New York: Springer-Verlag, pp. 251–260 (1980)

[44] PETRI, C.A.: *State-Transition Structures in Physics and in Computation.* International Journal of Theoretical Physics, Vol. 21, No. 12, pp. 979–992 (1982)

[45] PETRI, C.A.: *Concurrency Theory.* Petri Nets: Central Models and Their Properties, Advances in Petri Nets 1986, Part I, Proceedings of an Advanced Course, Bad Honnef, September 1986, Brauer, W.; Reisig, W.; Rozenberg, G. (eds.), Lecture Notes in Computer Science Vol.254, — Springer-Verlag, pp.4-24 (1987)

[46] PETRI, C.A.; SMITH, E.: *Concurrency and Continuity.* Lecture Notes in Computer Science Vol.266: Advances in Petri Nets 1987 / Rozenberg, G. (ed.) — Springer Verlag, pp.273-292 (1987)

[47] PLÜNNECKE, H.: *Schnitte in Halbordnungen.* Gesellschaft für Mathematik und Datenverarbeitung Bonn, ISF-Report 81.09 (Apr., 1981)

[48] PLÜNNECKE, H.: *K-Density, N-Density and Finiteness Properties.* Lecture Notes in Computer Science Vol. 188: Advances in Petri Nets 1984 / Rozenberg, G. (ed.) — Springer-Verlag, pp. 392–412 (1985)

[49] REISIG, W.: *Petri Nets — An Introduction.* Springer EATCS Monographs on Theoretical Computer Science / Brauer,W.; Rozenberg,G.; Salomaa,A. (eds.) (1985)

[50] REISIG, W.: *On the Semantics of Petri Nets.* Formal Models in Programming, IFIP 1985/ Neuhold, E.J.; Chroust, G. (eds.) — Elsevier Science Publishers B.V.: North Holland, pp.347-372 (1985)

[51] ROZENBERG, G.: *Processes of Elementary Net Systems.* Petri Nets: Central Models and Their Properties, Advances in Petri Nets 1986, Part I, Proceedings of an Advanced Course, Bad Honnef, September 1986, Brauer, W.; Reisig, W.; Rozenberg, G. (eds.), Lecture Notes in Computer Science Vol.254, — Springer-Verlag, pp.60-94 (1987)

[52] ROZENBERG, G.; THIAGARAJAN, P.S.: *Petri Nets: Basic Notions, Structure, Behaviour.* Lecture Notes in Computer Science, Vol. 224: Current Trends in Concurrency / de Bakker, J.W.; de Roever, W.P.; Rozenberg, G. (eds.) — Springer Verlag, pp. 585–668 (1986)

[53] SHIELDS, M.W.: *Adequate Path Expressions*. Lecture Notes in Computer Science Vol. 70: Semantics of Concurrent Computation / Kahn, G. (ed.) — Berlin: Springer-Verlag, pp. 249–265 (1979)

[54] SMITH, E.: *Zwei Bemerkungen über die Linearisierung von abzählbaren Halbordnungen*. Memorandum BEGRUND-40 der GMD (1985)

[55] STARKE, P.H.: *Petri-Netze*. VEB Deutscher Verlag der Wissenschaften, Berlin (1980) (in German).

[56] STARKE, P.H.: *Processes in Petri Nets*. Elektronische Informationsverarbeitung und Kybernetik (EIK) 17 8/9, pp.389-416 (1981).

[57] WINKOWSKI, J.: *Behaviours of Concurrent Systems*. Theoretical Computer Science 12, pp.39-60 (1980).

[58] WINSKEL, G.: *Events in Computation*. Ph.D. Thesis, University of Edinburgh (1980)

[59] WINSKEL, G.: *An Exercise in Processes with Infinite Pasts*. Informatik-Fachberichte 52: Application and Theory of Petri Nets/ Girault, C.; Reisig, W. (eds.) — Springer Verlag, pp.88-95 (1982)

[60] WINSKEL, G.: *Event Structures*. Petri Nets: Applications and Relationships to Other Models of Concurrency, Advances in Petri Nets 1986, Part II, Proceedings of an Advanced Course, Bad Honnef, September 1986, Brauer, W.; Reisig, W.; Rozenberg, G. (eds.), Lecture Notes in Computer Science Vol.255, — Springer-Verlag, pp.325-392 (1987)

Notation and Terminology

In this appendix we collect general notation and terminology which is used widely throughout the book but not related specifically to its contents. We assume the reader to be familiar with elementary mathematics, so that we do not define some notions (e.g., set, union, intersection, ...), nor state the basic facts (such as De Morgan's laws).

Logical Symbols

Suppose that A and B are logical expressions.

$A \wedge B$	A and B (logically).
$A \vee B$	logical or.
$A \Rightarrow B$	A implies B.
$A \Longleftrightarrow B$	A is equivalent to B.
$\neg A$	not A.
	Sometimes we indicate negation by a crossing symbol $/$, e.g. in $x \notin \ldots$ ('x is not an element of ...').
$\forall$	universal quantifier ('for all ...').
$\exists$	existential quantifier ('there exists ...').
$=$	equals.

Sets

Suppose that X and Y are sets.

$X \cap Y$	intersection of X and Y.
$X \cup Y$	union.
$X \uplus Y$	disjoint union.
$X \setminus Y$	set difference.
$X \subseteq Y$	X is included in Y.
$X \supseteq Y$	X contains Y.
$X \times Y$	Cartesian product.
$x \in X$	x is element of X.
$\{ \mid \}$	set brackets.
$\cap\{\ldots\}$	intersection quantifier.
$\cup\{\ldots\}$	union quantifier.
$\emptyset$	empty set.
$\mathbf{N}^+$	positive integers: $\mathbf{N}^+ = \{1, 2, 3, \ldots\}$.
$\mathbf{N}$	natural numbers: $\mathbf{N} = \{0, 1, 2, 3, \ldots\}$.
$\mathbf{Z}$	integers.
$\mathbf{Q}$	rationals.
$\mathbf{R}$	reals.
$|X|$	cardinality of X.

We write $|X| \in \mathbf{N}$ to denote the fact that X is finite, and $|X| \notin \mathbf{N}$ to denote the fact that X is not finite.

The Cartesian product of finitely many non-empty sets is non-empty. The axiom of choice, which is not derivable from the basic axioms of set theory, states that the Cartesian product of infinitely many non-empty sets is non-empty. The axiom of choice guarantees (i) that every li-set can be extended to a line and (ii) that every co-set can be extended to a cut. As a matter of fact, the axiom of choice is equivalent to (i) (which is known as the Kuratowski lemma) and is stronger than (ii).

Relations

Suppose that $\rho, \tau, \ldots$ are relations, i.e., subsets of the Cartesian product $X \times Y$ of two sets X and Y. We write $(x, y) \in \rho$ or $x \rho y$ to denote the fact that $x \in X$ and $y \in Y$ are in relation ρ.

$dom(\rho)$	domain of ρ: $dom(\rho) = \{x \mid x \in X \wedge \exists y \in Y: x \rho y\}$.
$cod(\rho)$	codomain of ρ: $cod(\rho) = \{y \mid y \in Y \wedge \exists x \in X: x \rho y\}$.
$\overline{\rho}$	complement of ρ: $(x, y) \in \overline{\rho}$ iff $(x, y) \notin \rho$.
ρ^{-1}	inverse of ρ (a relation on $Y \times X$): $(y, x) \in \rho^{-1}$ iff $(x, y) \in \rho$.
$\rho \circ \tau$	composition: if $\rho \subseteq X \times Y$ and $\tau \subseteq Y \times Z$ then $\rho \circ \tau \subseteq X \times Z$, defined by $(x, z) \in \rho \circ \tau$ iff $\exists y \in Y: (x, y) \in \rho \wedge (y, z) \in \tau$.
$\rho(x)$	for $x \in X$, $\rho(x) = \{y \in X \mid (x, y) \in \rho\}$.

Suppose that $\rho \subseteq X \times Y$, $X' \subseteq X$ and $Y' \subseteq Y$.

$\rho|_{X' \times Y'}$	restriction of ρ to $X' \times Y'$: $(x', y') \in \rho|_{X' \times Y'}$ iff $(x', y') \in \rho$.

Now suppose $X = Y$, i.e., $\rho \subseteq X \times X$.

$id|_X$	the identity relation on X: $(x, x') \in id|_X$ iff $x = x'$.
ρ^n	the n'th power of ρ (for $n \in \mathbf{N}$): $\begin{aligned} \rho^0 &= id|_X \\ \rho^{n+1} &= \rho \circ \rho^n \text{ for } n \in \mathbf{N}. \end{aligned}$
ρ^*	nonnegative iteration of ρ: $\rho^* = \bigcup_{n \in \mathbf{N}} \rho^n$.
ρ^+	positive iteration of ρ: $\rho^+ = \bigcup_{n \in \mathbf{N}^+} \rho^n$.

Functions

f is a partial function from X to Y means that f is a relation $f \subseteq X \times Y$ which satisfies
$$\forall x \in X \, \forall y, y' \in Y: (x, y) \in f \wedge (x, y') \in f \Rightarrow y = y'.$$

f is a total function (or just function) from X to Y means that f is a partial function such that, in addition, $X = dom(f)$. We write $f: X \to Y$ to denote f being a function from X to Y.

$f(x)$	function value.
injection	f is injective iff $\forall x, y \in X: f(x) = f(y) \Rightarrow x = y$.
surjection	f is surjective iff $Y = cod(f)$.
bijection	f is bijective iff it is both injective and surjective.

Arithmetic

Suppose that x and y are integers.

$x < y$	x is less than y.
$x > y$	x is greater than y.
$x \leq y$	x is less or equal to y.
$x \geq y$	x is greater or equal to y.
$+$	addition.
$-$	subtraction.
$*$	multiplication.
$-\infty, +\infty$	'numbers' adjoined to $\mathbf{Z}$ for the sake of convenience, with the property that $-\infty < z < +\infty$ for all $z \in \mathbf{Z}$.
$[x, y]$	closed interval between x and y: $[x, y] = \{z \in \mathbf{Z} \mid x \leq z \leq y\}$.

Suppose $X \subseteq \mathbf{Z}$ is a set of integers.

$\max(X)$	the maximal number in X (we use this only if we know that there is a maximal number in X).
$\min(X)$	the same with the minimal number in X.

Suppose I is a set.

$f: I \to \mathbf{Z}$	is an integer vector with index set I; if I is finite and $|I| = m$ then f can be represented as a vector with m components.
$f < f'$	ordering of vectors: if $f: I \to \mathbf{Z}$ and $f': I \to \mathbf{Z}$ then $f < f'$ iff $f \neq f'$ and $\forall x \in X: f(x) \leq f'(x)$.

Posets

A pair $(X, \prec)$ is a poset *iff* X is a set and $\prec$ (meaning 'precedes') is an irreflexive and transitive relation on $X \times X$; formally:

$\prec \subseteq X \times X$	$\prec$ is a relation on $X \times X$.
$id|_X \cap \prec = \emptyset$	$\prec$ is irreflexive.
$\prec^2 \subseteq \prec$	$\prec$ is transitive.

Now let $(X, \prec)$ be a fixed poset. We may define some derived concepts.

$\preceq$	precedes or equals: $\preceq \; = \; \prec \cup \, id|_X$.
$\succ$	succeeds: $\succ \; = \; \prec^{-1}$.
$\succeq$	succeeds or equals: $\succeq \; = \; \preceq^{-1}$.
$\prec\!\!\cdot$	immediately precedes: $\prec\!\!\cdot \; = \; \prec \setminus \prec^2$.

Further, let $A, A_1, A_2 \subseteq X$.

$\downarrow A$	down-set: $\downarrow A = \{x \in X \mid \exists y \in A : x \preceq y\}$.
$\uparrow A$	up-set: $\uparrow A = \{x \in X \mid \exists y \in A : y \preceq x\}$.
$[A_1, A_2]$	interval: $[A_1, A_2] = \{z \in X \mid \exists x \in A_1 \, \exists y \in A_2 : x \preceq z \preceq y\}$.
	(Remark: We shall need this definition only if A_1 and A_2 are co-sets (defined in Definition 2.1.1). If A_1 is a singleton, i.e., $A_1 = \{x\}$, then we write $[x, A_2]$ instead of $[\{x\}, A_2]$, and similarly for A_2. If $\mathbf{Z}$ is viewed as a poset with the usual ordering then the definition of interval just given is consistent with the one introduced in the section on Arithmetic.)
$Max(A)$:	set of maximal elements of A (can be empty): $Max(A) = \{x \in A \mid \neg \exists y \in A : x \prec y\}$.
$Min(A)$:	set of minimal elements of A (can be empty): $Min(A) = \{x \in A \mid \neg \exists y \in A : y \prec x\}$.
	(Remark: These definitions are consistent with the definitions of 'max, min' above; for the latter we always make sure that there is exactly one maximal (minimal) element, but for Max and Min there may be none, one, or more than one maximal (minimal) element(s).)

Index of Definitions